LASERS AND ELECTRO-OPTICS RESEARCH AND TECHNOLOGY

LASER TECHNOLOGY, APPLICATIONS AND FUTURE PROSPECTS

LASERS AND ELECTRO-OPTICS RESEARCH AND TECHNOLOGY

Additional books in this series can be found on Nova's website under the Series tab.

Additional E-books in this series can be found on Nova's website under the E-book tab.

PHYSICS RESEARCH AND TECHNOLOGY

Additional books in this series can be found on Nova's website under the Series tab.

Additional E-books in this series can be found on Nova's website under the E-book tab.

LASERS AND ELECTRO-OPTICS RESEARCH AND TECHNOLOGY

LASER TECHNOLOGY, APPLICATIONS AND FUTURE PROSPECTS

DETLEF EBERSBACH
AND
FELIX GOTTSCHALK
EDITORS

NOVINKA

New York

For permission to use material from this book please contact us:
Telephone 631-231-7269; Fax 631-231-8175
Web Site: http://www.novapublishers.com

LIBRARY OF CONGRESS CATALOGING-IN-PUBLICATION DATA

Laser technology, applications and future prospects / editors, Detlef Ebersbach and Felix Gottschalk.
pages cm.
Includes index.
ISBN: 978-1-62417-088-1 (soft cover)
1. Lasers--Industrial applications. I. Ebersbach, Detlef.
TA1677.L372 2013
621.36'6--dc23

2012037324

Published by Nova Science Publishers, Inc. † New York

CONTENTS

PREFACE

In this book, the authors present current research in laser technology applications and future prospects. Topics include the problems and solutions of ultra high power and intensity lasers; laser technologies applied for semiconductor HgCdTe and Chalcogenide thin films; prediction and analytical description of single laser tracks geometry and 316L stainless steel microstructure; and the laser induced transfer of organic materials.

Chapter 1 – The chapter will be devoted to one of the most rapidly growing area of laser physics, namely short pulse laser systems based on the chirped pulse amplification method (CPA).

The author will introduce it with a brief historical overview and the contemporary state of high power laser research. Two main ways of the reaching the high power will be discussed – systems with high energy, versus those with short pulse and relatively modest energy, where last one is a much more inexpensive way to reach most desirable applications of the high power laser pulses.

In the second part, the CPA principle as well as the main problems, which are raised on the way of reaching the high power and high intensity of the short laser pulse, will be discussed. Such restrictions as transverse parasitic generation and amplified spontaneous emission (ASE) in the large aperture final amplifiers, poor contrast of the high power output pulses, wave front distortion during amplification, compression and focusing and several others will be observed. Existing solutions of these problems and their weaknesses will be reviewed.

In the third part, the Hercules CPA laser system (Output power - 0.3PW, world records intensity - $2*10^{22}$ W/cm^2 and contrast -10^{11}) will be presented as an example of the classical CPA laser system. Several progressive ideas, which were developed and implemented into that system, allowing it to reach these parameters, will be described. Particularly, the extraction during pumping (EDP)-method, contrast cleaner with two non-linear crystals, method of the large aperture compressor alignment shortening of the laser pulse due to self phase modulation (SPM) in a bulk of the optical materials and several others will be discussed.

In conclusion, a list of the results as well as ideas for further development and perspectives of such kind of systems will be described.

Chapter 2 – CdHgTe-based semiconductors and structures as well as chalcogenide-based thin film active elements are of special interest for prospective applications in various fields of optoelectronics and information storage. Laser technologies made it possible to manufacture nanoscaled devices for far-infrared optoelectronics (CdHgTe and ZnCdHgTe solid solutions and thin films) and for optical information storage (Ag(Tl)Sb-containing thin films). The chapter reviews the latest results in fundamental and experimental research and reports unique data obtained by the authors. In particular, the growth technique and room-temperature electrical characteristics of ZhHgCdTe thin film based structures prepared by the pulse laser deposition and results from the investigations of laser technique-assisted structure engineering are presented. Applications of pulse laser deposition technology for producing long-life optical information storage media illustrate a wide perspective of manufacturing low-expensive and reliable active elements meeting needs of novel information storage.

Chapter 3 – Direct Laser Fabrication (DLF) is a process making possible the manufacture of functional parts directly from a command file injecting powder in a laser beam. The powder melting and solidification processes lead to the fabrication of this part layer by layer. In this study, deposition of 316L stainless steel powder on a steel substrate is carried out using a 700W fiber laser. Three values of each processing parameters (laser power P, scanning speed V and powder mass flow Q_m) are fixed and so 27 different experiments have been made and analyzed. The layer geometry is an important process characteristic and its mastery is essential to control the final part fabrication. Analytical relationships between the laser tracks geometrical characteristics (width, height, area, penetration depth) and the processing parameters are established. The proposed analytical relationships look like $y = a_0\left(P^{\alpha_1} Q_m^{\alpha_2} V^{\alpha_3}\right) + b_0$ where y is one of these geometrical characteristics.

Two kinds of models are explored to predict the clad geometrical form and characteristics. The first one is an analytical model in which the powder distribution in the feed jet is supposed to govern the laser clad geometry. It is then proved that the powder distribution in the jet can't determinate the final clad geometry. In the second one the general form of the clad cross section is supposed to be a disk due to the surface tension forces. Analytical relationships are established between the radius and the disk centre on one hand and the process parameters on the other hand. Comparisons between experimental observations and simulated geometries are very convincing.

Thermal study is carried out using the one source point method and the so called Green functions. The microstructure is experimentally observed, analyzed and explained with the thermal modelling through the solidification front velocity and the temperature gradients calculations.

Chapter 4 - The development of next-gen micro-electro-mechanical systems (MEMS) like sensors, actuators, biocompatible implants, photovoltaic cells, batteries, or electronics circuits is often limited by process equipment, its accuracy, performance and applicability. Newly, advanced materials (fragile organics semiconductors, large macromolecules, polymers or even complex composites) are also used for these devices. These problems can create almost insoluble question of processing technology choice.

One of the possible solutions lies in the area of laser technologies adapted for transfer and processing of organic materials. Conducting polymers (polypyrrole, polyaniline, polythiophene), pthalocyanines, acetylacetonates, carbohydrates, Teflon, carbon nanoparticles and their composites etc. were successfully deposited on many types of substrates with defined thickness and shape without molecular decomposition using Pulsed Laser Deposition (PLD), Matrix Assisted Pulsed Laser Evaporation (MAPLE), Laser InducedForward Transfer (LIFT) or Matrix Assisted Pulsed Laser Evaporation – Direct Write (MAPLE-DW). Although these technologies are still far from use in mass production, they have key and irreplaceable role in experimental or even in manufacturing of prototypes.

This contribution brings general overview of laser technologies, in particular those used in transfer and processing of organic materials supported by author´s team experience and experimental results. This contribution is aimed especially at: principals of special apparatuses for PLD, MAPLE, MAPLE-DW using high energy lasers (working in UV, VIS and IR spectrum, both pulsed and continual mode), theoretical analysis of the molecular transfer mechanisms and optimization of the process parameters for non-destructive deposition. Furthermore, results of the molecular transfer based on electron microscopy and infrared spectroscopy will be discussed. Finally, some applications of mentioned laser technologies will be proposed.

In: Laser Technology, Applications and Future Prospects
Editors: D. Ebersbach and F. Gottschalk
ISBN: 978-1-62417-088-1

Chapter 1

ULTRA HIGH POWER AND INTENSITY LASERS - PROBLEMS AND SOLUTIONS

Vladimir Chvykov[*]
Center for Ultrafast Optical Science, University of Michigan, Ann Arbor, MI, US

ABSTRACT

The chapter will be devoted to one of the most rapidly growing area of laser physics, namely short pulse laser systems based on the chirped pulse amplification method (CPA).

I will introduce it with a brief historical overview and the contemporary state of high power laser research. Two main ways of the reaching the high power will be discussed – systems with high energy, versus those with short pulse and relatively modest energy, where last one is a much more inexpensive way to reach most desirable applications of the high power laser pulses.

In the second part, the CPA principle as well as the main problems, which are raised on the way of reaching the high power and high intensity of the short laser pulse, will be discussed. Such restrictions as transverse parasitic generation and amplified spontaneous emission (ASE) in the large aperture final amplifiers, poor contrast of the high power output pulses, wave front distortion during amplification, compression and focusing and several others will be observed. Existing solutions of these problems and their weaknesses will be reviewed.

In the third part, the Hercules CPA laser system (Output power - 0.3PW, world records intensity - $2*10^{22}$ W/cm^2 and contrast -10^{11}) will be presented as an example of the classical CPA laser system [16]. Several progressive ideas, which were developed and implemented into that system, allowing it to reach these parameters, will be described. Particularly, the extraction during pumping (EDP)-method [46, 50, 51], contrast cleaner with two non-linear crystals [37], method of the large aperture compressor alignment [65] shortening of the laser pulse due to self phase modulation (SPM) in a bulk of the optical materials [75] and several others will be discussed.

In conclusion, a list of the results as well as ideas for further development and perspectives of such kind of systems will be described.

[*] Office (734) 936-4608; Lab (734) 936-0267.

INTRODUCTION

From the very beginning, when lasers were born in 1961 [1], people anchored their hopes on these devices as effective technological instruments and weapons. The main property required for these applications is a high output power. During that decade, such continuous wave (CW) -lasers as carbon dioxide-based [2,3] reached several kilowatt output power, and during the next decade the gas dynamic (chemical) lasers achieved high enough power (from tens of kW up to GW) [4] to be used in military, medicine (surgery) and industry for cutting, drilling, welding and other technological procedures. Recently the diode pumped solid states (thin ceramic disc and fiber) CW-lasers also reached very high output power (100's kW) [5] and became preferable due to the better beam parameters and compactness [6,7].

(a)

(b)

Figure 1. (a) - NIF Laser and Target Area Building; (b) - Laser Bay 2, one of NIF's two laser bays (from https://lasers.llnl.gov/about/nif/about.php).

Nevertheless, the easiest way to get extremely high powers is using the peak power of pulsed lasers, which require a much lower energy for the task. The situation became very promising when Q-switching [8] and mode-locking [9] regimes of laser operation were invented. This permitted to get a short enough laser pulse: nanosecond-level (10^{-9}s) at first and then consequently picosecond (10^{-12}s) and femtosecond (10^{-15}s) pulse durations. The lasers based on these technologies, and applying also master oscillator power amplifier (MOPA) configuration, achieved MW and GW-levels of power. The ability to reach these high powers allowed to look forward to the application of these systems in other new and very promising areas, such as charged particles acceleration [10] and inertial confinement of the fusion nuclear reaction (ICF) [11]. The consequent evolution of these laser systems resulted in the tremendous National Ignition Facility (NIF) laser in the United States [12] (Figure 1), and the Megajoule in Europe [13].

The prior, with its output energy in 192 laser beams about 4 MJ, is delivering to the target a power of up to 500 Terawatt. The laser occupies the area as large as three football fields (see Figure 1a) and consists of two laser bays, one of which is presented in Figure 1b.

The next multi order step in increasing laser powers was done more than twenty years after the Q-switch invention, when chirped pulse amplification (CPA) technology was suggested in the mid-eighties [14] (Figure 2).

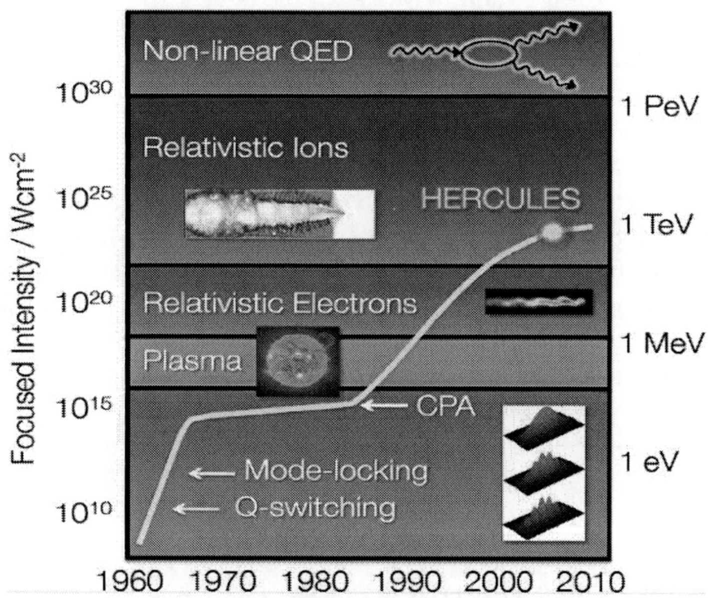

Figure 2. History of Laser Output Power Density Growth (from http://www.engin.umich.edu/research/cuos).

The new concept becomes possible with this technology, when modest enough energy allowed to achieve significantly higher powers using very short (tens of fs) pulse duration. Exploiting CPA-method, the researchers now reached a record output power – 1.5 Petawatts (10^{15} W) [15] and an intensity of 2×10^{22} W/cm^2 [16].

CPA Laser Systems – A New Branch of Laser Physics

The main goal of the CPA method is the reduction of size of optical elements and amplifier crystals. If one will try to amplify the short laser pulse directly, undesirable nonlinear distortions of the pulse in the gain media and other optical elements will be reached very soon. This leads to self-focusing of the laser beam, and thus a growing intensity, which damages the gain media and the optical elements. Therefore one has to increase the beam diameter and the apertures of the amplifiers and others optics to avoid these problems, but the attempt to keep the intensity under the self-focusing threshold requires enormous apertures of the optical elements (e.g., several meters in diameter for petawatt output power).

Very bright and progressive idea of the CPA was adopted for the optical diapason in the mid-eighties. It gave an alternative way to increase pulse energy. Instead of enlarging the transverse beam size, increasing the pulse duration was suggested for further amplification, which was then followed by the pulse compression. Historically, first time this idea was applied in the microwave diapason for radar technology in the 60-s [17]. Very soon the stretching and compression of the laser pulse was suggested using different methods, among which were the use of compression in dispersive media [18], multilayer film interferometers [19, 20] and, perhaps the most productive and widely explored today, the stretching and compression with a pair of diffraction gratings [21]. The first time when the pulse, chirped in a dispersive media (1.4 km of optical single mode fiber), was amplified, with further compression by a pair of diffraction gratings, was done by Donna Strickland and Gerard Mourou in 1985 [14]. However, the distortion of the chirped pulse, introduced by the higher order dispersion of the optical fiber was difficult to compensate by the grating compressor, and thus the output pulse duration was limited to a few picoseconds.

Most successful realization of this idea (Martínez-type stretcher setup containing a telescope between the gratings [22]) is presented in Figure 3, where the stretcher and compressor were both built on the base of diffraction gratings. The main idea can be explained very easily. Because a short pulse possesses a wide spectrum due to the Heisenberg uncertainty principle, we can redirect different parts of this spectrum (different harmonics) along different optical paths via diffraction, and thus get a different group delay for each. If in the stretcher, harmonics with longer wavelengths were directed along the shorter optical paths (positive or normal dispersion), then the compressor has to accomplish the reverse (negative or anomalous dispersion). Since the diffraction gratings are used both in the stretcher and the compressor, the functions of stretching and compression will be close to each other and the higher order dispersion can be compensated up to 4th order. This set up is then able to achieve pulse durations as short as 10's of femtosecond, if the Ti: Sapphire oscillators [23] with a few fs mode-locked pulse and amplifiers with a large bandwidth of fluorescence spectra are adopted. Two types of the amplifiers with these properties have been presented: the laser amplifiers [24], and optical parametrical amplifier (OPA) [25].

Simplified layout of the classical CPA laser system is presented in Figure 3. Here the short pulse from the mode-locked oscillator passing through the stretcher is expanded usually by 4-5 orders. Further the pulse with low fluence is amplified in the chain of amplifiers by several orders of magnitude, and is then compressed back to the short duration.

Exploiting this technology, researchers developed laser systems with petawatt-level output power but, as any other technology, CPA has some boundaries on the continued improvement of its parameters, the foremost of which is the limitation on the output energy.

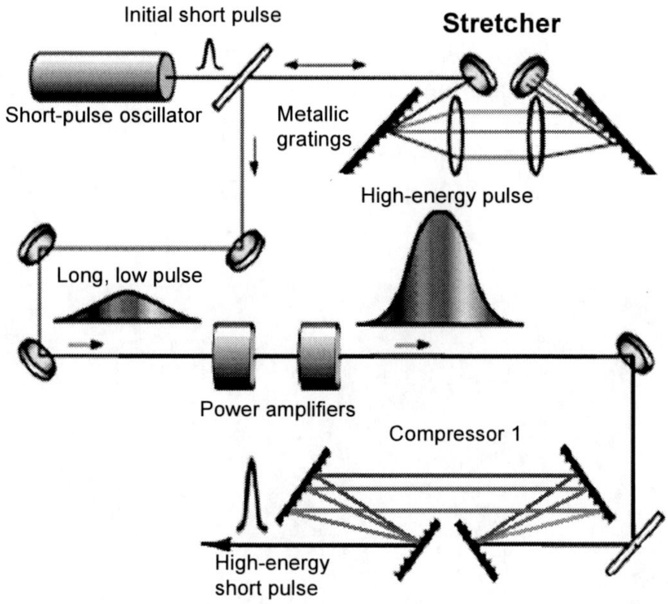

Figure 3. Simplified layout of the CPA laser system (from http://www.llnl.gov/str/Petawatt.html).

Limitation on the Output Energy

As mentioned above, recently several ultrahigh intensity CPA and, Optical Parametric Chirped Pulse Amplification (OPCPA) laser systems have reached petawatt output powers with the energy as low as several tens of Joules [26,15], setting the next milestone at tens or even hundreds of petawatts for the next five to ten years [27,32]. However, for that new generation of the ultra-high power lasers, the modest energy will no longer be enough, and the kJ-level has to be reached. The modern ability to stretch pulse duration is restricted at the few orders by the existing grating technology, so further energy increase requires once again the enlargement of amplifier apertures due to the damage threshold. The two candidates for the final amplifier in such systems are the optical parametrical amplifier and the laser amplifier. Both of them possess their own advantages and shortcomings. OPA has such attractive properties as the gain bandwidths large enough to support light pulses shorter than 10fs; “Optical cooling,” which leads to the potential of a high-repetition-rate; large high-quality nonlinear amplifier crystals; high pulse contrast due to the absence of amplified spontaneous emission (ASE) outside the pump pulse duration. At the same time there are severe requirements on the pump beam quality, such as limitation on the spatial fluctuations in amplitude; short pump pulses (few ps), in order to attain large bandwidth and high energy; the precise matching in space and time of the pump and the signal pulses, with a ps accuracy; and so on.

The laser amplifiers are free of most of these restrictions, but possess problems of their own, which include severe losses due to the transverse ASE (TASE), and the transverse parasitic generation (TPG). ASE as well as TPG leads to significant depletion of the inverted population, and thus the stored energy. The ASE from the pumped Ti: Sapphire crystal before and after the threshold of the parasitic generation is demonstrated in Figure 4 [26]. Because of this, enlarging the aperture of the amplifier fails to be an unlimited means of obtaining more output energy under the constraints imposed by the damage threshold limit. Therefore, in this case, the main limitation that arises on the path toward ultrahigh output power and intensity is the restriction on the pumping and extraction energy imposed by TASE and TPG within the booster amplifier volume. As a result, the suppression of parasitic generation is a very important task that has to be solved for the next generation of the ultrahigh power laser systems.

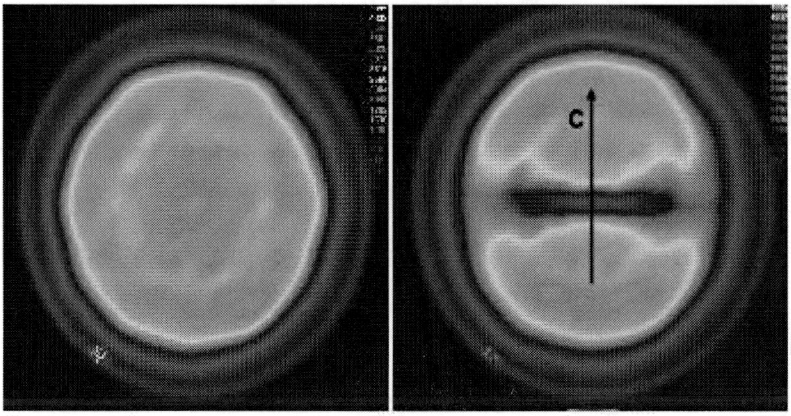

Figure 4. ASE from the pumped Ti: Sapphire crystal before and after threshold of parasitic generation.

Requirements on the Output Pulse Contrast

The other problem requiring an adequate solution is the necessity to have a good contrast of the output pulse. Remarkable progress in the CPA technology has recently allowed for laser intensities as high as 10^{22} W/cm^2 [28,16]. Such ultrahigh intensities create exciting opportunities for experiments in the ultra relativistic and radiation dominated regime [29] of laser-matter interactions. However, the laser contrast, defined as the ratio of the peak intensity of the pulse to the prepulse intensity, must be at least 11 orders of magnitude to restrict destructive preplasma dynamics. This estimate is valid for transparent targets with a high damage threshold [30], otherwise even higher contrast is required. If the contrast is not sufficient, a preplasma is created by the prepulse before the arrival of the main pulse, thus significantly complicating the experiments and depleting the effective intensity.

The Pulse Front Tilt (PFT) Problem

Growing output energy of the CPA-lasers [32, 31, 15] leads to the necessity to increase the diameter of the output beam in the compressor to avoid damage of the last grating. One of the problems arising from larger grating size is the requirement on the quality of the compression of the subsequently focused laser pulse. Therefore the larger gratings have to be aligned with a better accuracy. It is well known that non-collinearity of the grating pairs could produce a PFT of the output laser pulse. The tilt problems for pulses with large diameters were discussed in [33], where considerable increase of the pulse duration was demonstrated when focusing a tilted pulse. For example 10^{-4} radian non-collinearity of the pair of compression gratings, for a beam of 10" diameter could lead to the PFT of 2 * 10^{-4} radian and thus increase pulse-duration in focus by the factor of 5 for a 30 fs pulse.

Wavefront Distortion

The propagation of the amplified pulse trough optical elements and amplifiers of the laser system distorts its wavefront, and thus the generation of ultraintense light, which is simultaneously characterized by a high spatial accuracy, becomes a formidable proposition. On the other hand, to produce an ultrahigh intensity, it is more effective to obtain a tight focus than to further increase the laser power. However, tighter focusing is readily affected by aberrations, so the method of correction without including the aberrations of a focusing optic [34] will not achieve a good result. Therefore the compensation of wavefront aberrations during propagation through all systems and focusing is very desirable.

Gain Narrowing during Amplification

Finally, pulse duration of high-power CPA lasers is strongly limited by gain-narrowing in the gain medium (to around 30 fs for 100 TW-PW lasers based on Ti:Sapphire). Different approaches were entertained to overcome this limitation. Among them, there is the gain-narrowing control by introduction of the thin film etalon into the regenerative amplifier cavity. Impressive results were reached with this technique (pulse duration – 16 fs with spectral width - 72 nm) [35], but this method still suffers some restrictions, such as a finite possibility of the broadening of the spectra due to the fluorescence spectrum limitation of the amplifier crystal, further gain narrowing in the buster amplifiers, as well as the limitation of the spectral transmission of the grating compressor.

The application of concept of optical-parametrical chirped pulse amplification (OPCPA) for broadband pulse amplification allowed to reach a pulse duration of around 10 fs [36], but extremely severe requirements on the parameters of the pump laser restrict output energy only to a few millijoules. This way, an adequate method for shortening high energy pulses has not yet been found.

Practical Realization and Bottle-Neck Problems Solutions

Hercules laser, which was built in University of Michigan recently, will be presented as one of the most successful realization of CPA laser system [15] (Figure 5a). Several world record parameters such as intensity - $2*10^{22}$ W/cm^2 and contrast -10^{11} were reached and multiple battle-neck problems were solved during development of this laser. Layout of this system is presented on Figure 5b.

(a)

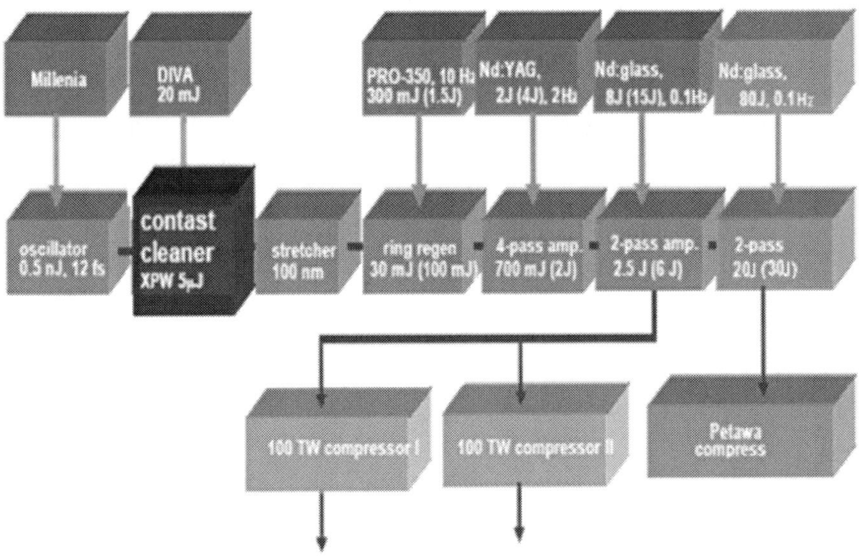

(b)

Figure 5. Hercules laser system: (a)- on The Laser Focus World cover; (b) – laser layout.

HERCULES laser design is based on CPA with cleaning of ASE- noise after the first amplifier. Output pulse of the short pulse oscillator (12 fs-pulse width, Femtolasers) is preamplified in the two-pass amplifier to the microjoule energy level. ASE added by the two-pass amplifier is removed by the cleaner based on cross-polarized-wave generation [37, 38]. Clean microjoule energy pulse is stretched to ~0.5 ns by the stretcher based on a modified mirror-in-grating design [39]. The whole laser is designed by a ray-tracing to be fifth-order-dispersion-limited over 100 nm bandwidth. High-energy regenerative amplifier [40] and cryogenically cooled 4- pass amplifier bring the pulse energy to a joule-level with nearly diffraction-limited beam quality (Figure 6 a, b).

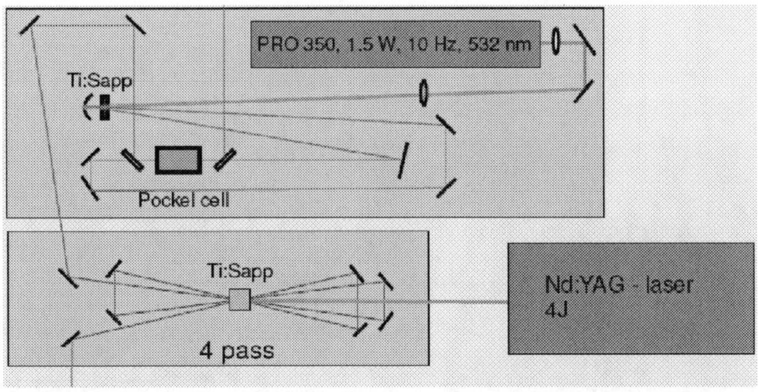

(a)

(b)

Figure 6. Regenerative amplifier and 4- pass amplifier: (a) – layout; (b) – cryogenically cooled chamber.

Two sequential 2-pass-Ti:sapphire amplifiers of 1' and 2" beam diameter respectively raise the output energy to a value approaching 20 J (Figure 7 a). We designed our own frequency-doubled Nd:glass pump laser [41] for pumping of the final two amplifiers of the HERCULES laser. The pump laser has two stages of amplification (Figure 7 b).

The frequency-doubled output of the first stage is used for pumping of 1"-diameter T:sapphire amplifier, while the unconverted infrared light is injected into the second stage of the pump laser for further amplification. The frequency- doubled output of the second stage is used for pumping of the booster (2"- diameter) amplifier of the HERCULES laser. The pump laser has a quasi-flat-top beam profile that was achieved at 0.1 Hz repetition rate by relay imaging and thermally-introduced birefringence compensation. The booster two-pass amplifier uses 11-cm-diameter Ti:sapphire crystal (Figure 7 c). Only a portion of this crystal is used to amplify 2" - diameter output beam of the HERCULES laser (Figure 8).

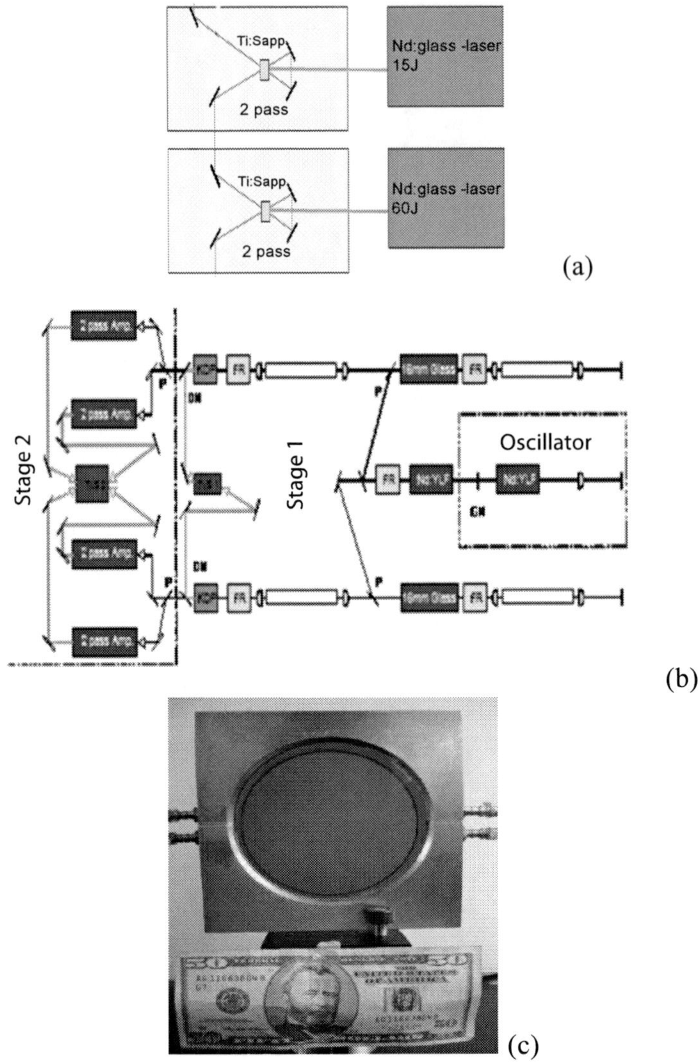

Figure 7. Frequency-doubled Nd:glass pump laser for pumping of the final two amplifiers: (a) – layout of the final two amplifiers; (b) - layout of the pump laser; (c) – crystal of the final amplifier.

Figure 8. The photo of the final amplifier during shot.

In order to suppress parasitic oscillations the side surface of the crystal is covered with a thin layer of index-matching thermoplastic coating (Cargille Laboratories, Inc.) doped with organic dye absorbing at 800 nm. Absorbing dye (comparing to ink used in single shot lasers [42, 43]) has an advantage of large absorption-cross-section and therefore allows using a relatively thin (a fraction of a millimeter) layer of the coating which is important for high repetition rate operation because of poor thermal conductivity of the thermoplastic.

Output beam profile (Figure 9 a) is quasi-flat-top as a result of using flat-top pump beams and of the image relaying of the amplified beam through the whole laser chain. Output energy of 17 J corresponding to 300 TW power after compression has been reached so far (Figure 9 b). The pump energy for the booster Ti:sapphire -amplifier (2”-diameter) is controlled by changing the pumping level of the oscillator of the pump laser. Because the same oscillator provides seeding pulse for both stages of the pump laser, changing the oscillator energy changes the pump energy for the last two Ti:sapphire amplifiers. It means that not only the pumping energy of the booster amplifier changes but the input energy changes as well if the oscillator energy is changed.

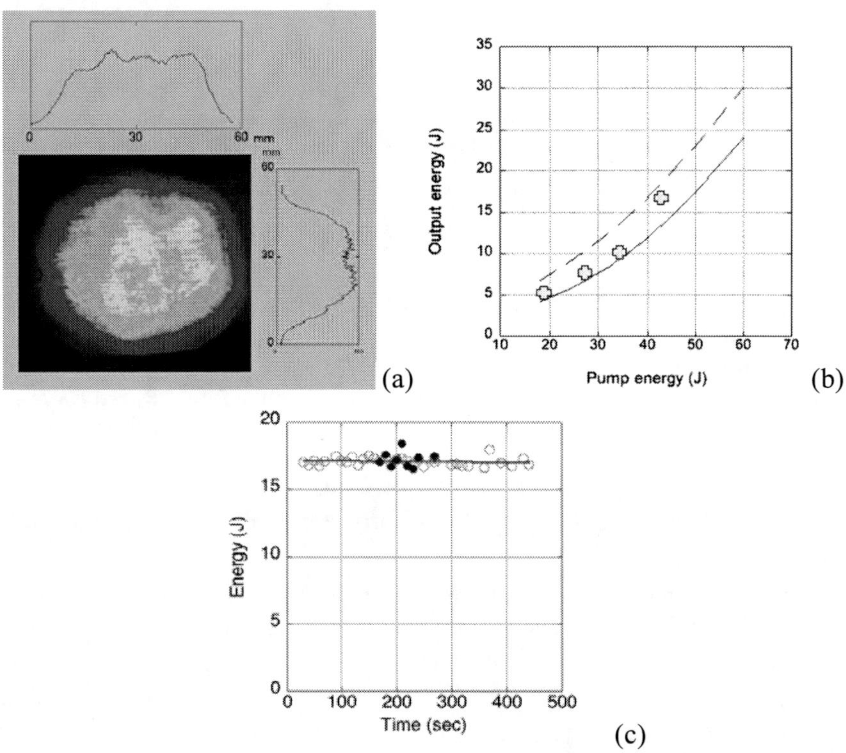

Figure 9. Measurements at the output of the booster amplifier. (a). Output beam profile of the HERCULES laser; (b). Output energy of the booster amplifier in dependence on the pump energy, crosses- experimental data, solid line-Frantz-Nodvic calculations for 1.7 J input energy, dashed line- for 3J input energy; (c). Output energy measured at 0.1 Hz for two shot series separated by 6 hours of continuous operation (shots represented by filled circles occurred 6 hours later then open circles shots).

As a result, in predicting the performance of the booster amplifier we calculate several Frantz-Nodvic curves, corresponding to varying input energy. Two of them, corresponding to input energy 1.7 J and 3 J are shown in Figure 9 b. Laser short-term and long-term stability is demonstrated at Figure 9 c , where the output energy measured at 0.1 Hz repetition rate is shown for two shot series (open circles and filled circles) separated by 6 hours.

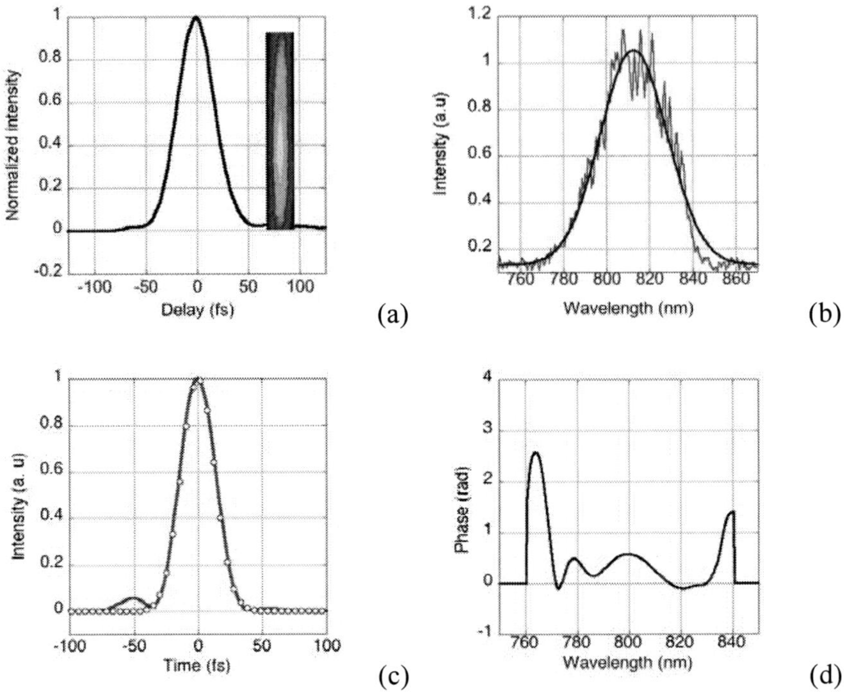

Figure 10. Compressed-pulse measurements. (a). Autocorrelation of 300 TW pulse, the experimental autocorrelation picture (insert) demonstrates that there is no amplitude front tilt or other spatial variations of the pulse arrival time; (b). Output spectrum: red curve - experimental data, blue curve - Gaussian fit (FWHM=37nm); (c). Intensity profile constructed by the SPIDER (red curve), Gaussian-fit-curve of 30.4fs - FWHM (blue-points); (d) Spectral phase reconstructed by the SPIDER.

The output pulse is compressed in a 4-grating compressor [42] to ~30 fs. The compressor is based on two 42x21cm-size and two 22x16.5cm-size -1200 l/mm-gold-coated holographic gratings (Jobin Yvon).

Because the beam size in the compressor is rather large (6”-diameter) achromatic lenses are used in the final relays to prevent spatially varying group delay across the beam. The autocorrelator that is sensitive to spatial variation of the group delay (autocorrelator with inversion [44]) is used to control this effect. The pulse width is measured at full energy using beam leak-through a mirror by two methods: autocorrelator with inversion- to ensure that there is no spatially varying pulse delay, and a single-shot spectral interferometry for direct electric field reconstruction (SPIDER) which was not sensitive to spatial variation of delay but was able to provide phase information for intensity reconstruction. The results of the measurements are shown in Figures 10 a-d. The intensity profile measured by SPIDER method (Figure 10 c) is well approximated by Gaussian-shape curve of 30.4 fs FWHM (with the exception of small pre-pulse at the foot of the pulse).

The shot-to-shot variation of FWHM is within 10%. The experimental spectrum profile (red line in Figure 10 b) is closely fitted by the Gaussian shape curve (blue line) of 37 nm FWHM. The pulse-widthband width product (~0.5) is close to the value of 0.44 for a transform-limited ~30 fs Gaussian pulse of 37 nm bandwidth that is also corroborated by a relatively flat spectral phase (Figure 10 d) at wavelengths where there is enough spectral intensity for reliable reconstruction by the SPIDER. The final amplifier added no more than 0.2 to the estimated B-integral value of the laser chain. This value is too low to influence the compression that is further evidenced by excellent quality of the compressed pulse (Figures 10 a, c).

Beam spatial quality is kept close to the diffraction limit (phase distortions $<\lambda$) by using of high-energy regenerative amplifier [40], which raised the energy to tens of millijoule level while maintaining diffraction limited quality of the beam and cryogenic cooling of the Ti:sapphire crystal for the subsequent 4-pass amplifier [37] (Figure 6 a, b).

Although final amplifiers are only water-cooled, the thermal effects in them are minimal as the average absorbed pump power for them is quite modest (<5W, they work at 0.1Hz repetition rate).

The beam is down-collimated by the all-reflective telescope after the compressor to 4”-diameter and is send to the interaction chamber where it is focused by the parabolic mirror.

A deformable mirror (4”-diameter, 177 actuators, made by Xinetics), controlled by the signal from a wavefront sensor located after the f/1 parabolic mirror (at low energy) compensates the aberrations of the parabolic mirror, astigmatism of the telescope and the residual aberrations of the laser beam. The focal distribution is characterized by using the method that we developed in [28,45].

And as result the HERCULES-laser demonstrated first time the power to 300 TW at high repetition rate, the highest intensity (~2 10^{22} W/cm^2) and highest contrast (~ 10^{11}). This intensity can be raised to 0.5 10^{23} W/cm^2 by using f/0.6 paraboloid (as was did in [27]) opening the radiation dominated regime of electron – light interaction for experimental studies.

Now let’s focus more detailed on the solutions of the several important problems, which were listed in the previous part and were solved during development of the Hercules – laser.

Suppression of Transverse ASE and Parasitic Generation

The conventional preventive procedure against parasitic generation is to reduce the reflectivity of the side (cylindrical) wall of the gain crystals by grinding, sandblasting and/or coating them with an index-matched absorptive polymer layer or liquids. However, the difficulty to introduce exact index matching restricts the diameter of the pump area to ~6 cm corresponding to an extracted energy to around 30 J from Ti:sapphire with the existing absorbers [15,26]. Attempts to enlarge the aperture of the amplifiers or to increase the pump fluence can lead to severe parasitic generation and fail to increase the extracted energy.

Moreover, if one will find method to reduce reflections down to zero, the losses remain still incredible big for amplifiers with the large aperture [46]. The uniform luminescence on the left picture on the Figure 4 should not delude us. Calculations reveal additional restriction for storing and extracting energy in this case. This restriction is even stronger than parasitic lasing because threshold for the latter can be increased with the development of the new index matched materials for absorbers. At the same time, the ASE in the transverse direction necessarily increases with the aperture size, which also reduces the maximum stored part of the pump energy.

We can estimate total volume of TASE radiated out from the crystal during its pumping. This calculation can be simplified assuming uniform distribution of density of population inversion in the transverse direction, which is realistic for the existing conditions due to the super-Gaussian profile of the pump beam. The differential equation governing the temporal dependence of population inversion is the first order nonhomogeneous equation:

$$\frac{dn}{dt} + an + W_s(n,t) = W_p \tag{1}$$

Where n – density of population inversion, a=1/τ, with τ – lifetime of laser transition, $W_s(n,t)$ – loss rate due to stimulated emission, W_p – pump rate.

If we split pump duration to *i*- time steps, which are, shorter than time required for significant changing of the population inversion, we can assume that W_s is constant during each time step. Therefore equation becomes linear (with initial conditions: $t_0=0$, $n_0=0$, $W_s^0=0$) and has well known analytical solution:

$$n_i = n_{i-1}e^{-t_i/\tau} + \tau\left(W_p - W_s^{i-1}\right)\left(1 - e^{-\frac{t_i}{\tau}}\right) \tag{2}$$

Now we have to find the value of W_s^i for each step of integration. For that, we will integrate ASE radiated during each time-step. Main part of the ASE passes through the sidewall of the crystal due to significantly higher gain, therefore for simplicity we disregard the other parts.

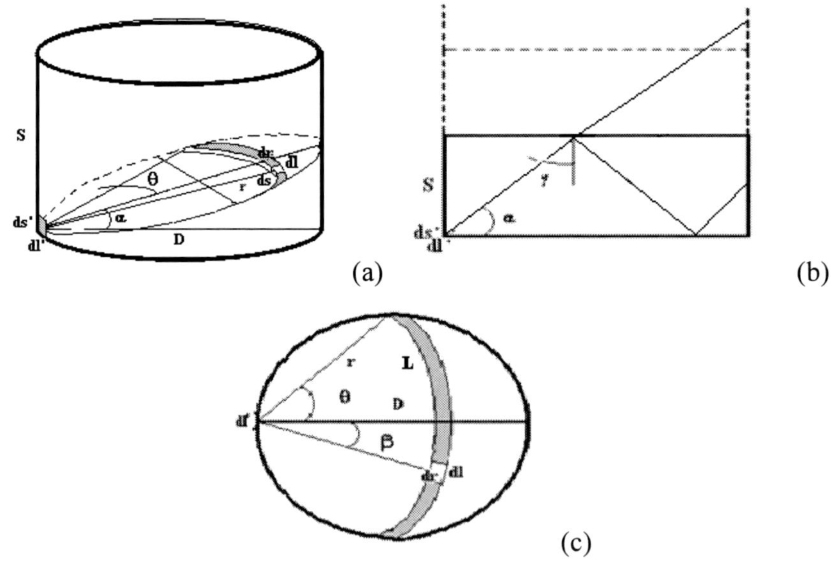

Figure 11. Amplifier crystal (picture for calculation of transverse ASE through side wall). (a) – total view; (b) side view; (c) – view from the top.

For integration we will take the element of crystal volume *dldrds* (Figure 11a), find fluence of ASE from this element through element of side surface *ds'dl'* and integrate over the whole volume.

$$\mathrm{dF_i} = \mathrm{d}l'\mathrm{d}s' \iiint \frac{h\gamma_e n' e^{\sigma n r} \cos\beta \cos\alpha}{4\pi r^2} \mathrm{dl\,dr\,ds} \quad (3)$$

Where dF_i – fluence trough the element of side area $dl'ds'$, $h\gamma_e$- emitted photon energy, *n'* – total decay rate of population inversion of the volume element *dldrds.*

To simplify the integration, we will use the geometry of crystal. Let's split the crystal into thin disks as shown on Figure 11 a, find integrated fluence from arbitrary one, and then integrate the angle α, when α sweeps from α_t to - α_t ($\alpha_t = (\pi/2)-\gamma$, where γ- critical angle of total internal reflection). This allowed taking in account ASE experienced total internal reflection from the flat surfaces of the crystal, as demonstrated on Figure 11 b.

The distribution of the population inversion in a longitudinal direction can be written as next equation when crystal pumped from both side: $n(s) = -aP_0\,(e^{as} + e^{a(S-s)})$, where α-coefficient of absorption of the pump light, P_0 – pump flax from one side, S – crystal sickness, s = *r sin α*, and crystal for purpose of calculation can be considered like virtual multilayer crystal (see Figure 11b).

For integration through thin disc we will draw circular arc L with the radius r from the element *dl'* and take the layer *dr* (Figure 11 c). The gain for ASE flux from each elements of this layer through the element *dl'* will be constant (equidistant). Therefore by replacing in (3), $dl = r*d\beta$,

$$r = \frac{D\cos\theta \cos\alpha}{\cos^2\theta \cos^2\alpha + \sin^2\theta};\ dr = \frac{-D\sin\theta \cos\alpha(\cos^2\theta \sin^2\alpha+1)}{(\cos^2\theta \cos^2\alpha + \sin^2\theta)^2} d\theta;\ ds = r*d\alpha$$

and taking into account that $\int_0^L \cos\beta\, dl = \int_{-\theta}^{\theta} \mathrm{r}\cos\beta\, d\beta = 2r\sin\theta$ we have the following integral for total ASE through the side surface, which can be evaluated numerically:

$$F_i = \frac{h\gamma_e S D^2}{2} \int_{-\alpha_t}^{\alpha_t} \int_0^{\frac{\pi}{2}} \frac{n' \mathrm{G} \cos^2\alpha \sin^2\theta(\cos^2\theta \sin^2\alpha+1)}{(\cos^2\theta \cos^2\alpha + \sin^2\theta)^2} d\theta\, da\,,$$

where $\mathrm{G} = \exp\left(\frac{\sigma n_{av} D\cos\theta \cos\alpha}{\cos^2\theta \cos^2\alpha + \sin^2\theta}\right)$; n_{av}- the average population inversion along r. Therefore $W^i{}_s = \frac{2F_i}{h\gamma_e S\pi D^2}$. Making this integration iteratively for each time-step throughout the pump duration we will get time dependence of *F(t)* and hence the losses during pumping process.

Figure 12a shows the evolution of normalized fluorescence of the crystals vs. pumping time when pumped by 100ns-pulse for different crystal apertures.

As seen from the plot, ASE grows dramatically after a certain time of pumping, even for 10 cm – crystal and soon becomes equal to the pumping energy. This means that further pumping is useless because all additional energy will be irradiated out of the crystal as ASE. These critical points of anomalous ASE (APs) are moving to the beginning of pumping process with the growing crystal diameter. No more than 50 -30% of the pump energy can be stored in the crystals with aperture of 15 -20cm.

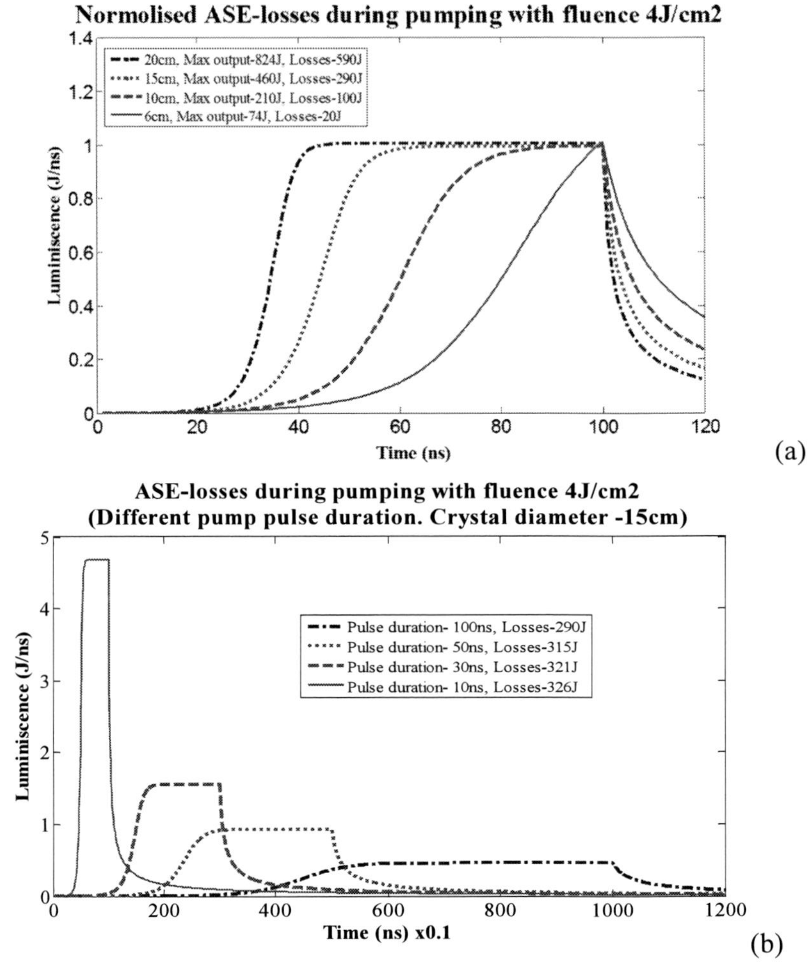

Figure 12. The fluorescence evolution of the crystals vs. pumping time. (a) - normalized ASE for different crystal apertures; (b) - ASE for different pump pulse durations (15cm crystal diameter).

Shortening of the pump pulse duration does not help to reduce the losses, at least until pulse duration becomes shorter than the time-length of the light distribution through the crystal in transverse direction. This is about several hundred ps, and thus such a pump would be useless due to the very low damage threshold. Figure 12 b shows crystal fluorescence during pumping for different pump pulse durations for 15cm crystal diameter. As seen from picture, the losses slightly grow with reduction of the pump pulse duration.

In order to reduce a transverse gain of the crystal near parallel surfaces the method of the double-pass pumping of a lightly-doped crystal was suggested in [47]. We calculate the losses due ASE for differently doped crystals (coefficient of pump absorption- α - 0.1 and 0.01) with different diameters. Results presented on the Figure 13a. As seen from the graphs, this method is able to reduce losses for crystals with relatively small diameters. When aperture is growing the losses increase in spite to the low doping and even become higher than the ones with the same aperture but the higher doping.

The matter of fact is, that there are two transverse modes: the generation near two working parallel surfaces due to the maximal population inversion [48] and the z-pass between these surfaces due to total internal reflection [49] (Figure 13 b). The latter surpasses the former in large aperture crystals with a high aspect ratio and respectively low crystal doping because higher gain for second mode. Moreover, because the deeper penetration of pump, the bigger part of the crystal volume involves in ASE generation process, therefore ASE reaches AP later but with the higher rate.

Assuming this discussion we have to conclude that the situation for laser amplifiers with the large aperture looks dramatically, however, our calculation also shows that these losses can be significantly reduced using EDP-amplifiers [50, 51], which are capable of helping with the TASE problem as well as with the parasitic generation.

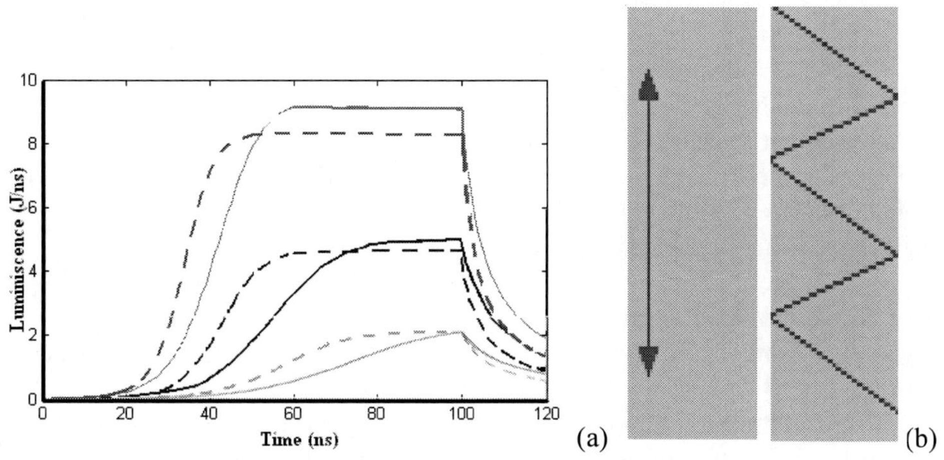

Figure 13. (a) –ASE for differently doped crystals with different diameters; (b) - two transverse modes.

Extraction during Pumping (EDP)

To overcome restrictions discussed above we suggest deviating from the conventional method of pumping and amplifying in multi-pass amplifiers, which is based on the energy stored in the upper laser level prior to the arrival of the first pass of the input pulse [52,53]. By continuing to pump after the arrival of the amplified pulse, we are able to forestall TASE and parasitic lasing and increase the extracted energy. In other words, the energy extracted during one pass of the amplified pulse through the crystal could be restored by pumping up to AP or the parasitic lasing threshold before the next pass [50]. The EDP process requires an extended pump pulse duration ranging from tens to hundreds of nanoseconds, or several delayed pulses. This allows sufficient time for proper pumping between passes, avoiding problems with temporal jitter and allowing increased total pump fluence due to the longer pump pulse duration. This approach was shown to double the output flux above the parasitic lasing limit in our experiments. The method has also been applied on several Ti: sapphire booster amplifiers of petawatt scale laser systems and has allowed output energies more than 60 J and reach record power 1.5 pettawatt from single channel [15,26].

EDP-Amplifiers Optimization

Here we will discuss the optimization of EDP for presently available large aperture Ti:sapphire amplifiers. The Frantz–Nodvik solution for the 1-D photon transport equation [53] is broadly applied to describe amplified transmission of pulses through the multi-pass amplifiers. But, in contrast to conventional amplifiers, where this equation is applied iteratively with adjustment of small signal gain for each pass, the solution for EDP-amplifiers can be rewritten as a single equation:

$$F_{out} = F_s \ln\left\{1 + \left[\exp\left(\frac{F_{in}}{F_s}\right) - 1\right]\exp\left(N\frac{\nu_{em}}{\nu_p}\frac{F_p}{F_s}\right)\right\} \tag{4}$$

where N is the number of passes, ν_{em} is the emission frequency, ν_p is the pump frequency, F_{out} is the N-pass output fluence, F_s is the saturation fluence, F_{in} is the incident fluence, F_p is the initial pump fluence (before signal arrives). The main difference between the EDP and the common multi-pass amplification is the restoration of the population inversion, and hence of the small signal gain, for each pass.

Therefore to get the output flux after N-passes of amplification we simply introduce a factor of N into the small signal gain expression. One may easily prove this by introducing the output flux from any N-1 passes as F_{in} for N-th pass and following the usual iterative procedure.

The results of the output fluence calculations for 4-pass amplification with EDP and with conventional amplification for various pump fluences are presented in Figure 14a. From this graph one can draw several important conclusions.

First — EDP amplifiers possess the ability to produce significantly more energy (up to four times for four-pass amplifier) compared to regular ones with the same initial pump fluence.

Second — because the graphs asymptotically approach the value of four times of the initial pump flux multiplied by coefficient of quantum defect, the input fluence that saturates the amplifier is much higher than that for the regular case. This is clear from comparison of the two green or two blue curves in Figure 14a. The reason for this is that with growing input flux we have a higher extracted energy and so are able to add more pump fluence between passes. The process of amplification mimics the case of a high value of F_s.

Third — we, therefore, see that the input fluence for amplification with EDP has to be much higher to make them efficient. Setting a criteria for the required input flux at half of the asymptotic value of the output fluence, which for curve of F_p=1.3 J/cm^2 is 3.5 J/cm^2, and corresponds to F_{in} of about 0.15 J/cm^2. Note that a four-fold increase in the value of the coefficient of small signal gain does not result in a commensurate increase in the total required pump fluence because we only need to replenish the deficit in the population inversion after each pass. The rhombus marked yellow curve of Figure 14b demonstrates the dependence of the total pump fluence on input fluence. The efficiency of EDP amplifiers is only a few percent lower than that of conventional ones. This can be seen by comparing the two blue curves with crosses and circles. Output fluence is nearly the same provided that the total input fluences for the usual amplifier and the EDP-amplifier are equal (green curves with triangles and squares on Figure 14b), but the conventional amplifier is not able to be operated at the same initial pump fluence due to TASE or parasitic generation.

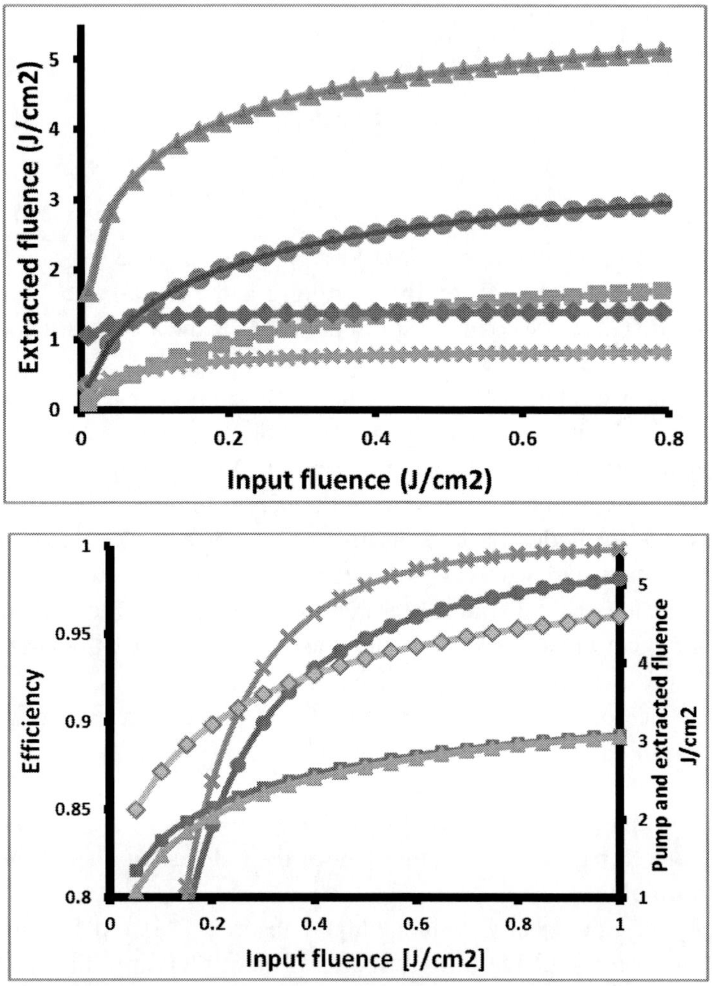

(a)

(b)

Figure 14. (a)-dependence of the output fluence of the 4-pass amplifier on the input fluence for several pump fluxes ($F_P = 2$ J/cm^2 - two green curves, triangles – EDP amplifier, rhombs – no EDP; $F_P = 1.3$ J/cm^2 – two blue curves, circles - EDP amplifier, crosses – no EDP; $F_P = 0.8$ J/cm^2 EDP-amplifier – purple squares. (b) – total pump and 4-pass extracted versus input fluence for 25 cm amplifier, yellow rhombus curve – pump of the EDP-amplifier; two green curves- extracted fluencies, triangles – EDP, squares - no EDP; efficiency of 4-pass amplifiers with $F_P = 0.8$ J/cm^2 - two blue curves, circles – EDP, crosses – no EDP.

It might now seem that the problem of TASE and TPG can be solved by the EDP method and so the extracted energy could be increased without bound by enlargement of the amplifier crystal aperture. However, this is not precisely so, as will be clear from the following discussion.

The key parameter for calculating the output fluence is the initial highest pump flux below the parasitic generation or AP. As was mention above, there are two transverse parasitic modes: the generation near two working parallel surfaces and the z-pass between them (Figure 13 b). The product of the pump fluence and the diameter of the pump area of the crystal can be explicitly written as:

$$D \times F_p = K \quad ,$$

where $K = \frac{\nu_p}{\nu_{em}} \times \ln G_{pt} \times \frac{zF_s}{\ln B}$ for gain near surface, and $K = \frac{\nu_p}{\nu_{em}} \times \ln G_{pt} \times \frac{zF_s}{2n}$ for z-pass,

(5)

where D is the pump area diameter, G_{pt} is the highest transverse gain that would be compensated by an index-matched absorptive coating, z is the crystal thickness, n is the index of refraction of the crystal or AR coating if present, and B is the absorption coefficient for the pump frequency.

The product of the crystal diameter and the pump fluence is constant for each value of the critical transverse gain or index matching of the absorber and crystal thickness, so that a crystal with a given aperture fixes the maximum pump fluence. Further, enlargement of the crystal aperture requires reducing the pump fluence. Moreover, because the complete matching is impossible for both polarizations of the transverse modes, the pump fluence will be limited for any crystals and absorbers. In Figure 15(a) curves showing the dependence of the pump fluence on the crystal diameter are presented. The purple rhombus-marked curve is an upper bound for 3 cm crystal thickness and was calculated for σ- polarization under the condition of total index matching for π- polarization.

Thus, the existence of an optimal diameter for the pump area can be demonstrated by introducing into formula (4), in place of F_{in} and F_p, the ratios E_{in}/A and K/D, respectively, where E_{in} is the total incident energy and A is the pump area. The relationship between the output energy and the diameter of the pump area, for different K values and incident energies, is presented in Figure 15(b). The area on the right of the red dashed line lies below the output fluence damage threshold.

As an example, for a practically available liquid absorber [26] and a crystal with 20 to 25 cm diameter [54], we can calculate the highest output energy for amplification with EDP. The maximum transverse gain this absorber is able to accommodate is 4000, which leads to a value of K= 32. From the blue rhombus curve in Figure 15(b), we find the output energy of ~800J, which corresponds to 50 J input energy. In reference [15], the authors demonstrate 60 J with EDP in the final amplifier, making 800J realistic output energy for the next stage of amplification.

In cases when losses due ASE reach AP before development of the parasitic generation, EDP-amplifiers also are able significantly suppress them.

Estimations demonstrate that four–pass EDP amplifier with 15 cm diameter pump area and 100 ns pulse duration can approach almost theoretical limit the extraction energy, namely 325J when pumped by 560J. Figure 16a demonstrates the fluorescence depends of time for that crystal of EDP-amplifier (compare with Figure 12). The timing between passes is close to optimum from the point of maximizing the extracted energy and so losses are only 17J, as opposed to the 290J for the conventional amplification.

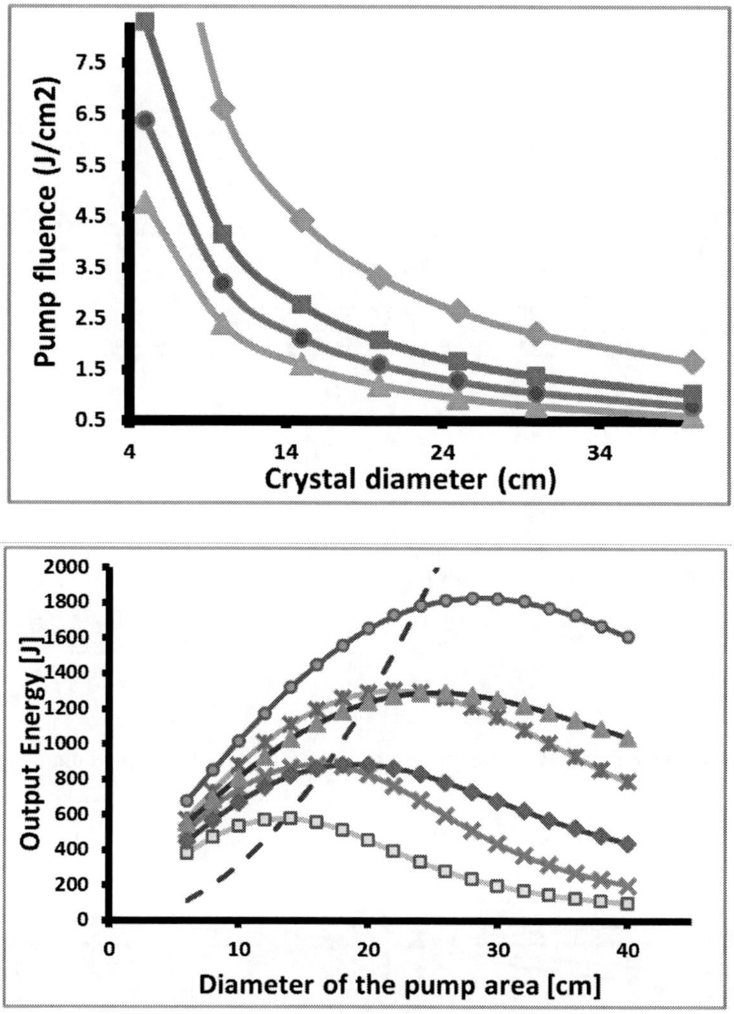

(a)

(b)

Figure 15. (a) - dependence of the pump fluence on the crystal diameter for different K-factors (K=24 – green triangles, K=32 – blue circles, K=42 – brown squares, K=66 boundary curve – purple rhombs); (b) - dependence of the output energy on diameter of pumped area for different input energies and K-factor (10 J input energy – K=32 yellow squares, K=42 blue crosses; 50 J – K=32 blue rhombs, K=42 green asterisks, 150 J – K=32 green triangles, K=42 purple circles); dashed red curve – damage threshold fluence.

Another example of the pumping the same crystal is presented on the Figure 16b. Delivering 400J of the pump energy in green, the EDP-amplifier will supply 230J. With the constant pump rate, this timing could be implemented with the appropriate geometry of the amplifier. One of the possible compact packed schemes is presented on Figure 17. As seen from Figure 16b the delays between 1st and 2nd passes or 2nd and 3rd are 12-15ns and between 3rd and 4th passes – 25ns which corresponded to presented geometry with bypass of the crystal between 3-d and 4-th pass.

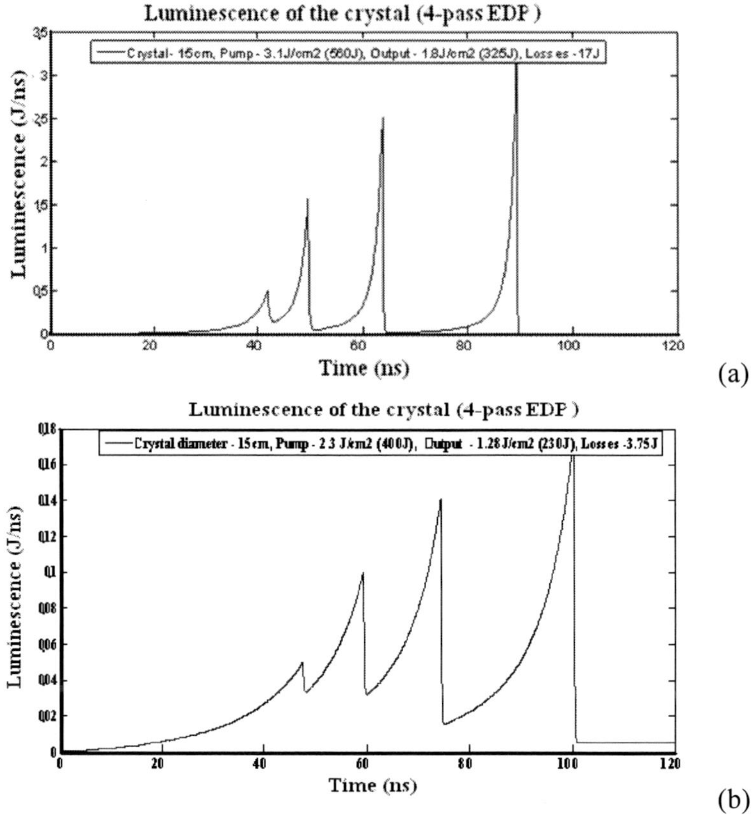

Figure 16. The fluorescence depends of time for crystal of EDP-amplifier. (a) – pump energy 560 J; (b) – pump energy 400 J.

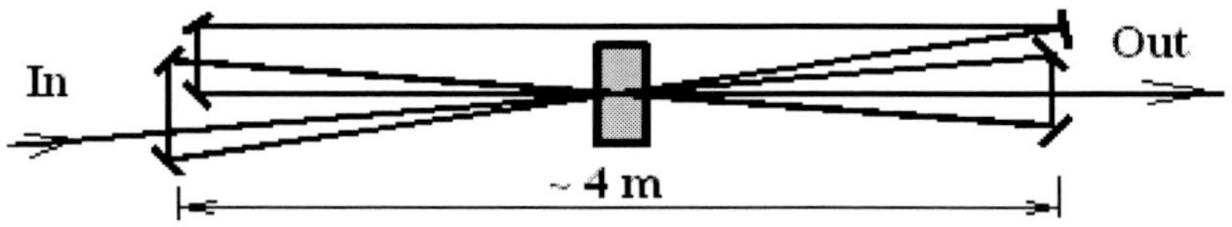

Figure 17. Schematic diagram of the 4-pass EDP-amplifier.

Summarizing this part, - our estimations reveal the ASE as a more strong restriction then transverse parasitic generation for increasing the aperture of Ti: sapphire amplifiers.

- EDP-amplifiers, when operated under optimal conditions, are capable to suppress the losses connected with transverse ASE and parasitic lasing and increase the extracted energy by up to four times with the good efficiency, as compared with conventional amplification of the same initial pump fluence.
- The existence of an optimal pump aperture for each given index-matched absorber and incident energy was shown.

- CPA laser systems with EDP-amplifiers recently approach record output power – 1.5 PW.
- With available index-matched liquid absorbers and large aperture Ti: Sapphire crystals, it is possible to approach kJ level extracted energy by exploiting EDP. With 60% compressor transmission efficiency and 10-15 fs pulse duration, 30 - 50 PW power level could be reached from a single channel.
- To reach this very high extracted energy, at least the last two amplification stages of the laser system, will have to implement EDP.

Contrast Cleaning

The two major types of laser prepulses are ASE and those caused by nonideal compensation for the total phase introduced through the CPA laser system. Fortunately, phaseconnected prepulses are generally less detrimental than ASE because the arrival time scale is of the order of tens of picoseconds before the peak of the pulse, resulting in insufficient time for the preplasma expansion to affect the experimental conditions compared with what occurs for plasma created by ASE (which has a nanosecond time scale).

Several pulse-cleaning techniques have been developed for ASE suppression. At the input of the cleaner the main laser pulse must be short because the stretched pulse overlaps the ASE, making discrimination against ASE within the stretched pulse impossible. Accordingly, all methods suggested for discrimination against ASE, such as the use of saturable absorbers [55, 56], polarization rotation [57,58], double CPA [59] cross-polarized-wave (XPW) generation [60,38], and plasma mirrors [61] can be classified in two major groups. The first group uses the amplified short pulse before stretching, and the second (plasma mirrors and frequency doubling) uses the pulse after compression. Of the methods using pulse amplification and cleaning before the stretcher, only the relatively simple saturable absorber technique has been practically implemented in a multiterawatt laser. The contrast improvement achievable with this method, however, is limited to 1–2 orders of magnitude. Contrast improvement to 10^{10} with double CPA was demonstrated in a model small-scale experiment; however, it has not yet been implemented in multiterawatt lasers because of complexity and stability issues. We demonstrate a 50 TW laser with a contrast of 10^{11} without the use of a double CPA system [37]. This result is based on two major developments.

First, the HERCULES laser contrast without a pulse cleaner is relatively high (10^8 compared with common values of 10^6–10^7) due to the use of a long cavity-ring-resonator regenerative amplifier that was developed [40] for HERCULES. This contrast level is achieved with nanojoule pulses injected into the resonator. Injecting a clean pulse (contrast more than 10^{11}) of microjoule energy, instead of nanojoule energy, will allow the pump fluence to be lowered to achieve the same output energy. This will result in a decrease of ASE generation accordingly. Second, we show that the complicated and expensive double CPA method that was previously used to acquire millijoule energy pulses is not necessary (at least up to intensities of the order of 10^{22} W/cm^2) because efficient pulse cleaning can be done at the microjoule energy level. The contrast was directly measured to be at least 10^{11} with a third-order autocorrelator. To our knowledge HERCULES is currently the highest-contrast multiterawatt laser system [37].

The XPW method of contrast improvement [61] has the advantages of simplicity and solid-state operation. For millijoule-energy pulses, however, a separate CPA system is necessary for the sole purpose of generating the input pulses for XPWs. We modified XPW generation for microjoule input energy pulses, which can be generated by direct amplification and does not require the use of an additional CPA.

The XPW method is modified in the following two ways to use it with microjoule pulses: first, we use two BaF_2 crystals in series to improve XPW conversion efficiency. Second, a programmable acousto-optical filter (Dazzler from Fastlite) was used to compensate dispersion of the systems optics and amplifier. As a result, the pulse duration was lowered to the value close to that at the oscillator output 12 fs, which increases the focused intensity on the BaF_2 crystal.

A simplified schematic of the HERCULES laser with a more detailed cleaner layout is shown in Figure 18. The oscillator output pulse 4 nJ, 12 fs is switched out by a Pockels cell with a 10 Hz repetition rate and is amplified in a two-pass collinear preamplifier to an energy of several microjoules. The same Pockels cell separates the input of the amplifier from the output.

The preamplifier is pumped from both sides with up to 20 mJ total energy from a frequency-doubled diode-pumped $Nd:YVO_4$ laser. A programmable acousto-optical filter compensates for the dispersion of the Pockels cell and the amplifier. The output pulse of the preamplifier is focused by a spherical mirror on two 2-mm-length BaF_2 crystals and is recollimated by a second spherical mirror. The output polarizer rejects the low-intensity prepulses and transmits the XPW generated in the BaF_2 crystal by the main pulse to the stretcher.

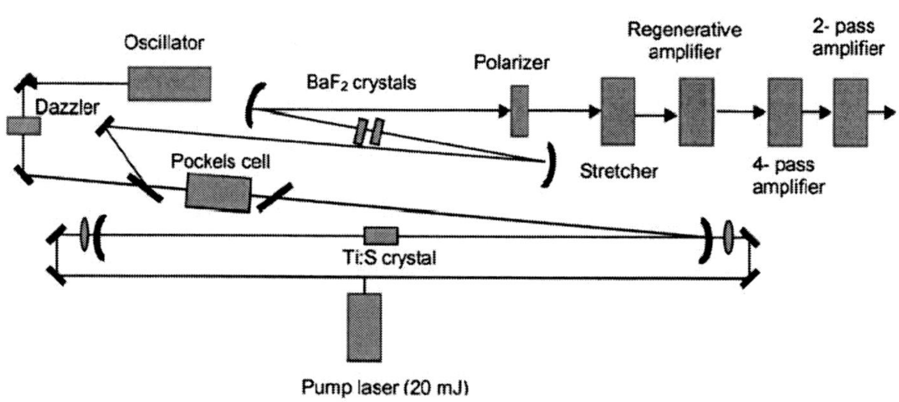

Figure 18. Simplified layout of the HERCULES laser.

The ring regenerative amplifier amplifies the pulse to 40 mJ. A cryogenically cooled 4-pass and a water-cooled 2-pass amplifier raise the output peak power to 50 TW after compression (energy of 1.5 J, pulse duration of 30 fs). Previous work using XPW generation demonstrated a laser contrast of 10^{10} with millijoule input pulses that were focused onto a 2-mm-long BaF_2 crystal with a 3-m-focal length lens. In our experiments, however, we focused our beam with a 20 cm radius of curvature mirror to provide the intensity necessary to generate a XPW with microjoule pulses. The Rayleigh range of the focused beam in this case becomes less than the crystal length, thus lowering the conversion efficiency. In addition, dispersion of the amplifier and the Pockels cell makes the pulse longer, which also lowers the intensity. These combined effects lower the measured conversion efficiency to a value of less than 1% in a single-crystal cleaner. We raised the value of the conversion efficiency to a value comparable to that demonstrated in [38] (10%) for millijoule-energy pulses and in [60] for microjouleenergy pulses by compensating dispersion with a Dazzler. To further increase the conversion efficiency two BaF_2 crystals were used. Use of two identical crystals is more efficient than a single crystal of double the length. This is due to the existence of an optimum angle (22.5°) between the laser polarization vector and the crystallographic axes of the crystal needed for efficient generation of the XPW. Using a second crystal, whose axes can be independently rotated to match the direction change of the polarization vector, helps to compensate for the polarization rotation in the first crystal and allows for higher conversion efficiency than in the single-crystal setup. An efficiency of 25% has been achieved with two BaF_2 crystals (Figure 19a). We measured the contrast (Figure 19b) at the laser output by using a third-order autocorrelator. The contrast was measured with and without the cleaner in place. The detection floor of the autocorrelator was $5*10^{-12}$. The signal peaks at the 10^{-10}–10^{-11} level are due to single photoelectrons. Note that the contrast is underestimated in Figure 19b because of the lowering of the main pulse amplitude at zero delay that is caused by the limited temporal resolution of the third-order autocorrelator 130 fs. It was difficult to perform contrast measurements using all HERCULES' amplifiers because of the lower repetition rate of the 2-pass amplifier 0.1 Hz_ than the 2 Hz repetition rate of the 4-pass amplifier and the 10 Hz rate for the regenerative amplifier. Accordingly, we measured the contrast without cleaning with only the regenerative amplifier engaged at 10 mJ output energy (black curve in Figure 19b).

Contrast with the cleaner was measured in two ways: at 2 Hz (with all amplifiers except the last one engaged, leading to 10 TW output power, red curve) and by plotting several points at 50 TW 1.5 J with all amplifiers engaged. It is clear from Figure 19 that the cleaner improves the ASE contrast by at least 3 orders of magnitude. The noise added to the output pulse by the two booster amplifiers is much less than the noise of the regenerative amplifier amplified by the booster amplifiers. This result follows from the next consideration. For a single-pass amplifier the ASE flux is [52]

$$I_{ASE} = I_S \frac{\Omega G_0}{4\sqrt{(lnG_0)}} \tag{6}$$

where I_S is the saturation flux, Ω is the solid angle, and G_0 is the small-signal gain of the amplifier. For a multipass or regenerative amplifier (typical gain of 2–3 per pass) most of the ASE is generated at the first couple of passes and is amplified on the subsequent passes until the gain is saturated by the main pulse. The contrast K is then

$$K = \left(\frac{I_{ASE1} G_0^{N_{sat}}}{I_{inj} G}\right)^{-1} \tag{7}$$

where I_{inj} is injection flux, I_{ASE1} is the ASE flux after the first pass, N_{sat} is the number of passes until the gain is saturated, and G is the total gain of the amplifier for the main pulse. It follows from Eqs. (6) and (7) that the noise generated directly by the booster amplifiers is much smaller than the amplified preamplifier noise for two reasons. First, the small-signal gain of the regenerative amplifier (10^9 without the cleaner and 10^6 with the cleaner) is much larger than the combined small-signal gain of the booster amplifiers $4*10^3$. Second, in contrast to the regenerative amplifier, where for the majority of the passes gain is not saturated, the gain of the booster amplifiers is always saturated and is heavily saturated at the end passes. Additional confirmation of the 3 orders of magnitude improvement of contrast in Figure 19 comes from the gain consideration. In HERCULES, laser material dispersion of the regenerative amplifier is used to zero the total fourth-order phase distortions, therefore the total number of passes through the regenerative amplifier (22) is defined by the amount of fourthorder phase and cannot be changed regardless of whether the cleaner is used. Because the output from the cleaner is much larger than the oscillator output, the pump energy of the regenerative amplifier was lowered from 360 to 200 mJ, causing a consequent lowering of the small-signal gain per pass from 2.7 to 1.75. A calculation using equations (6) and (7) and these numbers for small-signal gain per pass results in a contrast improvement of 1000, which is in good agreement with Figure 19b.

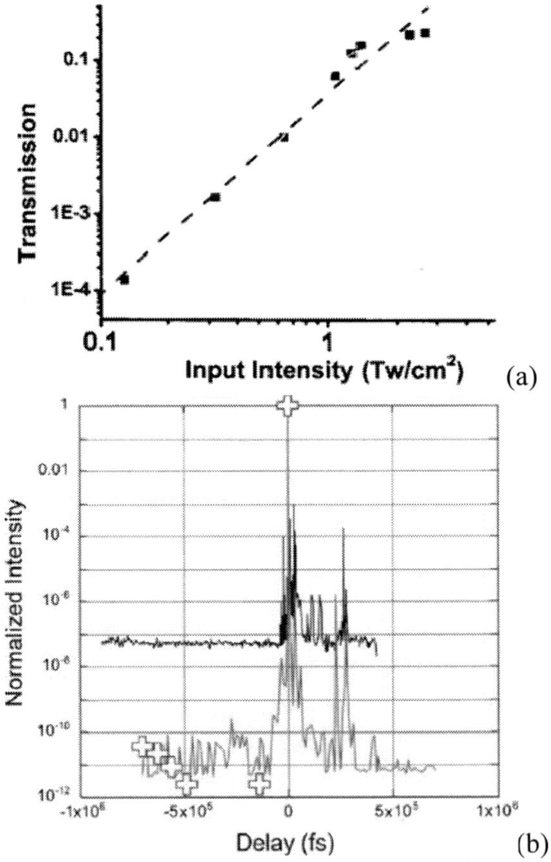

(a)

(b)

Figure 19. a- Dependence of the transmission of double-crystal XPW cleaner on intensity at the crystal; b - Third-order autocorrelation with (red curve, 10 TW power; crosses, 50 TW power) and without (black curve, regenerative amplifier only) cleaner. To get a contrast value, the intensity from the third-order autocorrelation has to be divided by 4, a ratio of the temporal resolution of the autocorrelator to the pulse width. The peaks of the red curve at the 10^{-10}–10^{-11} level are due to single photoelectrons, corresponding to approximately four photons (quantum efficiency of the photomultiplier photocathode 25%).

In conclusion, we have demonstrated that pulse cleaning based on XPW generation in two BaF_2 crystals yields a 10^{11} contrast ratio for a 50 TW laser system. Dispersion compensation and direct amplification of the oscillator output are necessary for efficient XPW generation using only microjoule energy. Two BaF_2 crystals placed in series increase the XPW conversion efficiency by a factor of 3 compared with a single crystal, yielding 3 orders of magnitude of contrast improvement. Further addition of the booster two-pass amplifier [16] and the double plasma-mirror after compressor [62] allowed to achieve contrast ratio at 10^{15} with output power 300TW.

THE QUALITY COMPRESSION OF THE LASER PULSE WITH THE LARGE GRATINGS COMPRESSOR

Several methods for alignment of compressor with large gratings were previously suggested. One of these methods utilizes direct control of the grating alignment by a single-shot autocorrelator [63], the other uses several types of interferometers [64]. Each of these techniques has some shortcomings. For example, the use of a single-shot autocorrelator leads to a multi-stage-procedure for alignment, while the conventional interferometers introduce additional optical elements and, with them, additional alignment steps and distortions.

We introduced a new, very simple, accurate and practical method for compressor alignment and demonstrated its advance in art for grating size adequate for use in compressors of Petawatt-scale lasers [65]. In our method of the compressor alignment the compressor gratings are used simultaneously as gratings of the diffraction interferometer. We mix the beams of different orders from different gratings and use interference patterns for indication of the grating collinearity.

Figure 20a shows a schematic diagram of the procedure. Beam 1 of the laser used for aligning, (normally HeNe laser) arrives on the grating 1 with the angle given by the equation $sin(\alpha) = \lambda/d$, where λ is the wavelength and d is the grating constant. The diffracted angle for first order beam (beam 2) in this case equals 0. The incident angle value is experimentally chosen by observing the second order diffraction beam (beam 3), which has to be exactly opposite to the path of the beam 1. Beam 2, reflected by grating 2, returns to grating 1 and experiences first order diffraction, resulting in beam 5, which, mixes with the specular reflection of beam 1 (0th order, beam 4). The interference pattern with good contrast can then be monitored on screen. It is then possible to align grating 2 relative to grating 1 by rotating it around X and Y axes and eliminating the fringes of the interference pattern.

In addition to parallelism, the groves of both gratings have to be aligned accurately. The groves (axis Z, Figure 20b) are aligned after gratings are made parallel (as described above) by directing beam 1 to grating 1 at Littrow angle. Beam 2 (0th -order) in this case arrives to the grating 2 after specular reflection at the same angle because of collinearity of the gratings. Interference pattern can be observed by the mixing reflection from grating 2 of beam 2 and retro reflection of beam 4 (beams 3 and 6) or the retro reflections of beams 1 and 2 (beams 4 and 5, 1st -order diffraction for Littrow angle) . Rotation of the grating 2 around the axes Z eliminates the interference fringes and aligns grooves of the gratings collinearly.

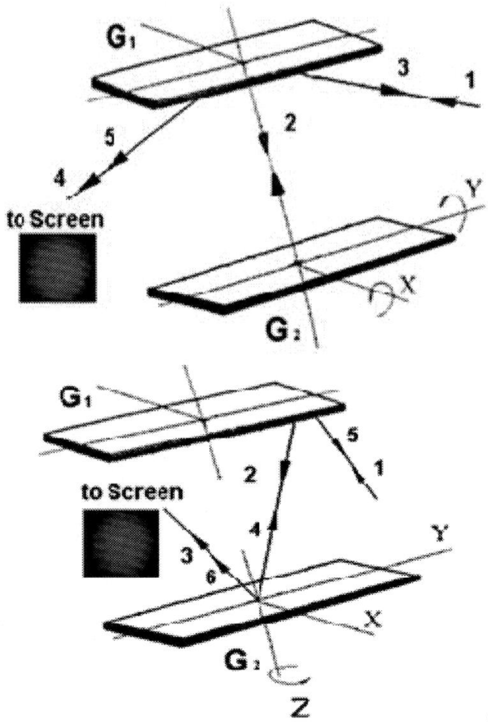

(a)

(b)

Figure 20. Schematic diagram for grating alignment: a – collinearity of the gratings, b – collinearity of the grooves.

Precision of this method is depends on the diameter of beam 1 and is limited only by the grating size. With the grating size of about 50 cm this can be done with accuracy of up to 10^{-6} radian.

In the “Hercules” laser system the 6” diameter output beam is compressed to 30 fs pulse duration. Figure 21a demonstrates the Hercules laser compressor. To keep the pulse duration at this value +10%, the precision of the grating alignment has to be not worse then 10^{-5} radian. For this, the beam of the HeNe laser was expanded to 2.5” diameter. Measurements of the pulse duration and PFT were done by a single shot autocorrelator with inversion. Figure 21b shows autocorrelation of 300 TW pulse. An experimental autocorrelation picture (insert) demonstrates that there is no PFT or other spatial variations of the pulse at arrival time.

Accuracy of the tilt measurements can be estimated using the following considerations. From the picture of the autocorrelation one can conclude that the tilt angle is no more then 10^{-2} radian. The ratio between autocorrelation tilt and PFT for small angles and the configuration of the autocorrelator used is about $6*10^{-2}$ [66]. Then, because the output beam from the compressor is down collimated for measurements by 100 times, the angle of the pulse front versus phase front could be estimated not worse then $0.6*10^{-5}$ radian.

We note in conclusion that pulse tilt angle is a not particularly convenient parameter of the pulse because it depends on the beam size. It is more practical and convenient to use the projection of the pulse front on the beam direction normalized to the pulse length, which is connected to the PFT through the formula *tg(a)d/(tc)*, where *t* is pulse duration, *c*- the speed of light, *a*- angle between the pulse and the phase front , and *d* – the beam diameter. This parameter could be measured directly from autocorrelation and is independent of the beam size.

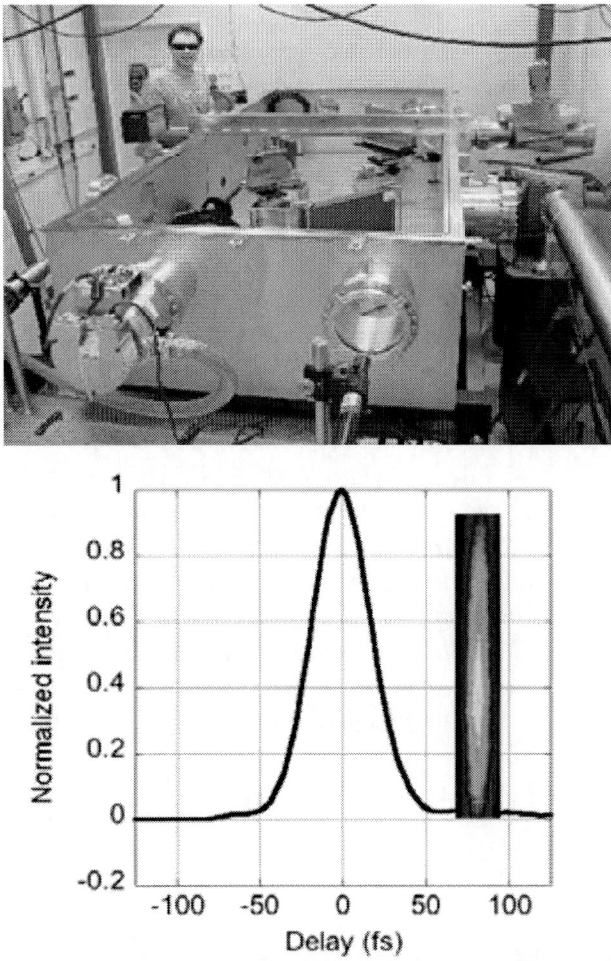

(a)

(b)

Figure 21. a - Hercules compressor; b - autocorrelation of 300 TW pulse.

WAVE FRONT CORRECTION

We developed a new method of the measuring and correction aberrations of the optical elements and amplifiers of the laser system as well as a large-numerical-aperture focusing optic [28,45]. The fact that we measure the phase after focusing leads us to an effective way to characterize the field. By use of a diffraction integral with measured phase and amplitude, the field, after focusing, can be calculated everywhere [67]. Such field calculations have occasionally been performed but with low accuracy because the aberrations of the focusing optics were not characterized [68]. The field calculation technique yields a more detailed analysis of the experimental results than spot size measurement or ionization [69].

The application of tight focusing and aberration correction is shown schematically in Figure 22. The 6.7 cm by 5.4 cm elliptical laser beam (attenuated regenerative amplifier beam) is focused by an *f/ 0.6* paraboloid (7.62-cm diameter, 5.08-cm effective focal length, 90±). The all Ti: Sapphire laser (Hercules) essentially consists of a regenerative amplifier (40 mJ) and a four-pass amplifier followed by a two-pass amplifier (2.5 J before and 1.4 J after the compressor). The power amplifiers exhibit minimal aberrations because the four-pass amplifier is cryogenically cooled and the two-pass amplifier is operated at a low repetition rate (0.1 Hz). The FWHM bandwidth at the regenerative amplifier level is 45 nm, and it becomes 43 nm at 30 TW. The autocorrelation trace of a 10-TW beam indicates a 30-fs pulse width that results from a Gaussian fit, where we have used a tilted pulse-front single-shot autocorrelator.

The wavefront aberrations are measured through an imaging system consisting of a f ield-curvature-corrected, apochromatic-infinity-corrected objective (403; N.A., 0.75) and an achromatic lens (*f* - 25 cm). In detail, the objective forms a virtual image of the deformable mirror (DM) near the focus, and this image is relayed onto a Shack–Hartmann wave-front sensor (Imagine Optic) by an achromatic lens. The sensor has a 5 mm by 5 mm capture area of 8-bit dynamic range with a 32 by 32 microlenslet array. The focal length and the diameter of each microlens are 5.6 mm and 153 μm, respectively. The depth of focus is thus 400 λ, so the sensor is insensitive to a 45-nm bandwidth. As the diffraction effect is negligible from the DM to the paraboloid, the paraboloidal aberrations can be linearly added to the laser beam aberrations. This linearity allows us to directly control without further complications the wave-front distortion that is due either to the laser beam or the paraboloid.

The DM was manufactured by Xinetics, Inc., and it consists of 96 actuators (lead magnesium niobate) controlling a 7.62-cm-diameter mirror surface by use of the electrostriction effect. The coating of the mirror (protected silver) can withstand as much as 300 mJ/cm^2 of energy density with 30-fs pulses. The maximum interactuator stroke is 2 mm.

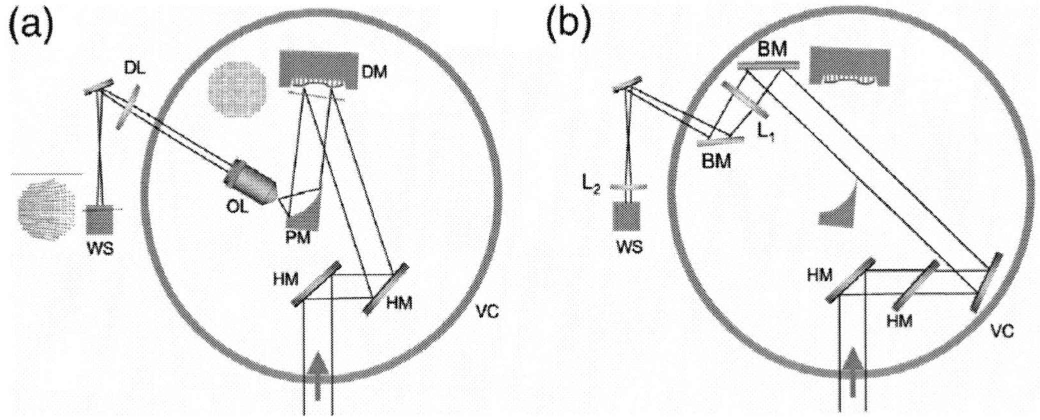

Figure 22. (a) Experimental setup for measuring and correcting laser beam and focusing optic aberrations by use of a regen amplifier beam: HM, high-ref lectivity mirror; DM, deformable mirror; PM, *f*/0.6 paraboloid; OL, inf initycorrected apochromatic–planatic objective with a N.A. of 0.75, 403; DL, doublet lens; WS, Shack–Hartmann wave-front sensor; VC, vacuum chamber. (b) Setup for measuring the phase of a high-energy beam compared with that of a reference beam (regen beam): BM, blank mirror; L_1, lens, f_1 -50 cm; L_2, lens, f_2 -4 cm.

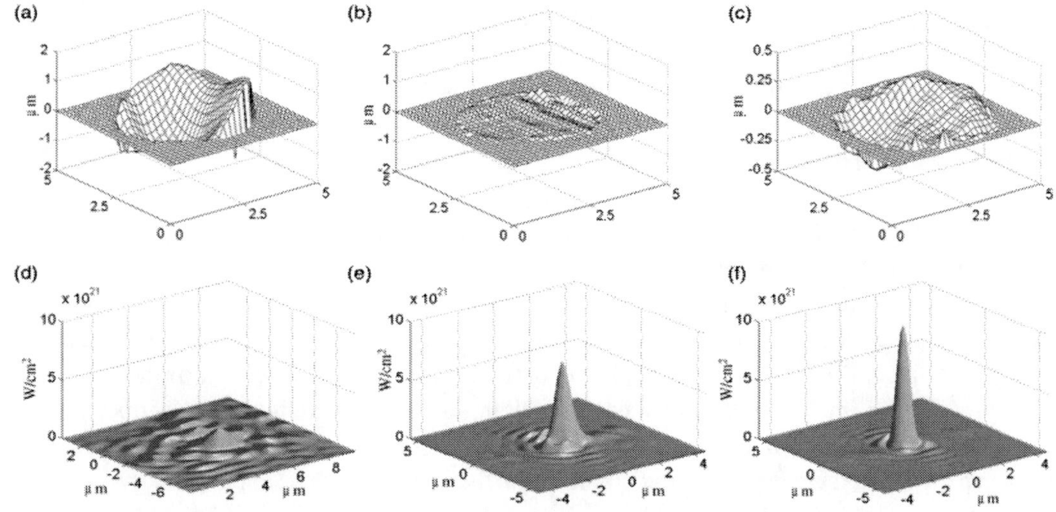

Figure 23. (a)–(c) Measured wave fronts and (d)–(f) calculated focused intensities. (a) Wave front measured after focusing by a regen beam with no correction. rms, 0.936λ; peak-to-valley distance, 4.247λ. (b) Corrected wave-front after focusing by a regen beam. rms, 0.066λ; peak-to-valley distance, 0.618λ. (c) Differential wave front at 45 TW versus regen beam. rms, 0.154λ; peak-to-valley distance, 0.657λ. (d) Calculated focal intensity with no correction at full power (45 TW). I_{peak} - 10^{21} W/cm^2. (e) Calculated focal intensity after correction of the aberrations of the regen beam and the paraboloid. I_{peak} - 6.9 * 10^{21} W/cm^2. (f) Calculated focal intensity with the correction of the aberrations from power amplifiers. *I*peak - 10^{22} W/cm^2.

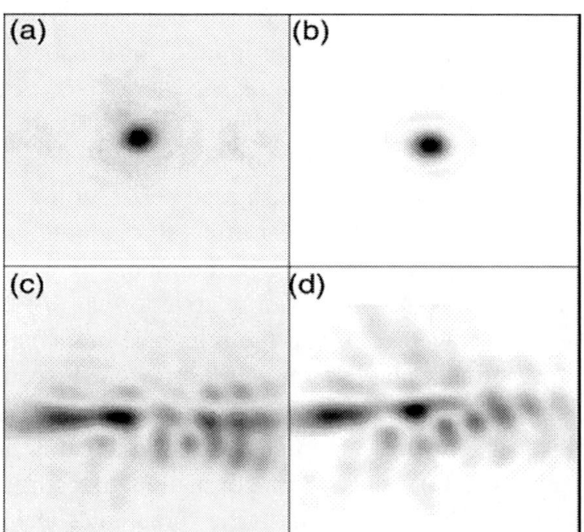

Figure 24. (b), (d) Comparison of focal spot calculation and (a), (c) measurement with a 12-bit CCD camera at the regen level. (c), (d) before correction; (a), (b) after correction. (b), (d) Calculated by use of measured phase and amplitude profile after focusing.

The wave-front measurements before and after correction by the regenerative amplifier (regen) beam are shown in Figs. 23a and b. The root-mean-square wave-front quality improved from 0.936λ to 0.066λ. A 12-bit camera image of the focal spot after wave-front correction is shown in Figure 24a. The FWHM spot size of the corrected focal spot is 0.8 μm, and the spatial resolution of the image is 40 pixels/mm. The main lobe of the corrected spot in the 12-bit image contains approximately 60% of the total energy, whereas the calculated spot contains 80% (Figure 24b). The discrepancy can be attributed to the sensitivity to defocusing error of encircled energy of the experimental picture. In this tight focusing configuration, 1.5-mm defocusing results in 25% reduction in the encircled energy.

The wave-front data shown are subject to two sources of error: statistical and measurement errors. The first is due to uncontrollable, random f luctuations of the wave front such as air turbulence and thermal drift in the amplifiers. The measurement errors are due to the finite sampling of the Shack–Hartmann sensor or to imperfect imaging in the wave-front measurement. The measurement errors in the imaging system (Figure 22a) are estimated to be 0.02λ rms. This estimation is based on the rms measurement of the wave front through a similar imaging system by use of a light source from a single-mode fiber tip. The sampling error is negligible because the analysis of a two-dimensional Fourier transform of the beam profile suggests a small number of high spatial frequency terms that are undetectable by the Shack–Hartmann sensor. Also, by inserting the attenuation filters we introduce nonnegligible first-order wave-front tilts as well as negligible higher-order terms (,0.015λ rms). The wave-front tilt errors have to be corrected with an external pointing system. We calculate the statistical rms values, however, by averaging the rms value of each wave front relative to the mean wave-front shape using 10–20 shots. The statistical rms values were 0.04λ at the regen and four-pass amplifier and 0.06λ at the two-pass amplifier level (30 TW). The overall limit of the wave-front correction is determined by the statistical rms value, i.e., the shot-to-shot stability of the wave front. As these errors are small, the measured focal spot distributions match the calculated results well (Figure 24).

After successfully achieving a wavelength-limited focal spot size with a low-energy reference beam (regen beam), we measured the higher-energy phase before focusing because we cannot use the objective owing to concerns about damage (Figure 22b). At this step, only the phase of a high-energy beam relative to that of the reference beam needs to be measured. To perform this measurement we took the transmitted part of a high-energy beam (45 TW) through a high-reflectivity mirror. The nonlinear effect through the reflector is negligible. In the relative measurement mode the aberrations in the imaging system were subtracted out. Figure 23c shows the phase of the 45-TW beam relative to that of the regen beam. The rms value is 0.154λ. As the high-power beam and the reference beam follow the same path and traverse the same optics, the total input field at high energy can be constructed by addition of the residual reference phase (Figure 23b) and the differential phase. However, direct addition of the two fields is not possible owing to the field redistribution after parabolic reflection (see the spot diagrams in Figure 22a). Thus a transformation of coordinates is necessary for describing the field before and after reflection off the paraboloid. The transformation is given by

$$x_0 = \frac{2f(-p\cos\varphi + m\sin\varphi)}{(1 + p\sin\varphi + m\cos[\![(\varphi)]\!]}, \quad y_0 = \frac{2f(-q)}{(1 + p\sin\varphi + m\cos[\![(\varphi)]\!]}.$$

where f is the parent focal length of a paraboloid and p, q, and m $[=(1-p^2-q^2)^{1/2}]$ are direction cosines viewed from the focus and can be considered coordinates of the reflected field. (X_0, Y_0) are coordinates of the field before reflection across the beam near the paraboloid. Φ is the mean incidence angle of the focusing beam (74.3^0). After transformation of the differential phase in terms of the reflected field coordinates, the two phases were summed for diffraction calculation. The amplitude was also measured and taken into account. The diffraction integral used is derived for an off-axis paraboloid from the Stratton–Chu formula [67] with a far-field approximation. The far-field formulas satisfy all four of Maxwell's equations in vacuum. To calculate intensity, we consider only monochromatic diffraction; the finite pulse duration effect becomes significant only in the subcycle pulse regime [70]. Figures 23d , e show the calculated focused intensity distribution without any corrections and with correction of the aberrations in the regen beam and the paraboloid at 45 TW, and the peak intensities are 1 * 10^{21} and (6.9 6+/-0.7)* 10^{21} W/cm^2, respectively. The peak intensity can be further increased to (1.0+/- 0.1)* 10^{22} W/cm^2 by correction of the differential phase at high energy (Figure 23f). The correction for the high-energy beam can be achieved by addition of the differential phase to the wave-front sensor measurements of the low-energy correction mode. This correction is possible only because of shot-to-shot wave-front stability. Thus the correction of the paraboloid and the low- and high-energy laser beam aberrations can be achieved through accurate measurement of the wave front.

We have demonstrated that by focusing a 45-TW laser beam to a spot size of 0.8 μm a record peak intensity of 1 *10^{22} W/cm^2 can be reached. To obtain such a tight focus with an excellent Strehl ratio of 0.9 we use an *f* /0.6 paraboloid and correct it by a 96-actuator deformable mirror.

After laser upgrading to 300TW a deformable mirror (4"-diameter, 177 actuators, made by Xinetics), controlled by the signal from a wavefront sensor located after the f/1 parabolic mirror compensated the aberrations of the parabolic mirror, astigmatism of the telescope and the residual aberrations of the laser beam. The focal distribution was characterized by using the method described above, that allowed to achieve record intensity 2*10^{22} W/cm^2 with the focal spot 1.3 μm.

Applications for such high intensity concentrated within one wavelength are the efficient generation of isolated attosecond pulses and attosecond electron bunches [71]. The direct wave-front correction is applicable only in a laser system that produces a stable shot-to-shot wave front.

We have demonstrated a technique based on accurate wave-front measurement that can provide amplitude and phase information anywhere in the focal volume with high spatial resolution. We believe that the full characterization of a high-intensity beam will provide a new tool for accurately simulating experiments in this intensity regime. We believe that it will be possible to demonstrate an intensity of 10^{23} W/cm^2 following an upgrade of Hercules laser system.

Two Stage Compression

The secondary compression method based on spectral broadening using self-phase modulation (SPM) with the further recompression of the chirped pulse (Figure 25) looks very promising. This is a method for obtaining very short (around 10 fs) amplified pulse however, is not suited for the highest powers because it uses SPM with combination of spatial filtering in near field [72] in order to provide for spatially uniform phase due to the axial uniform part of the Gaussian distributed intensity. This method suffers also the limitations of incident energy due pinhole damage and looses resulting in transmission efficiency of just a few percents and output energy of several microjoules. To reduce looses the spatial filtering was used in far-field either with a waveguide [73] or bulk medium [74]. Homogenization of the spectra broadening through whole beam due to diffraction allowed in this case to improve transmission efficiency to up to 40% and reach output energy around hundreds microjoules. But the both approaches were also restricted to only a millijoule-level output energy due to burning of the spatial filter diaphragm or hollow fiber.

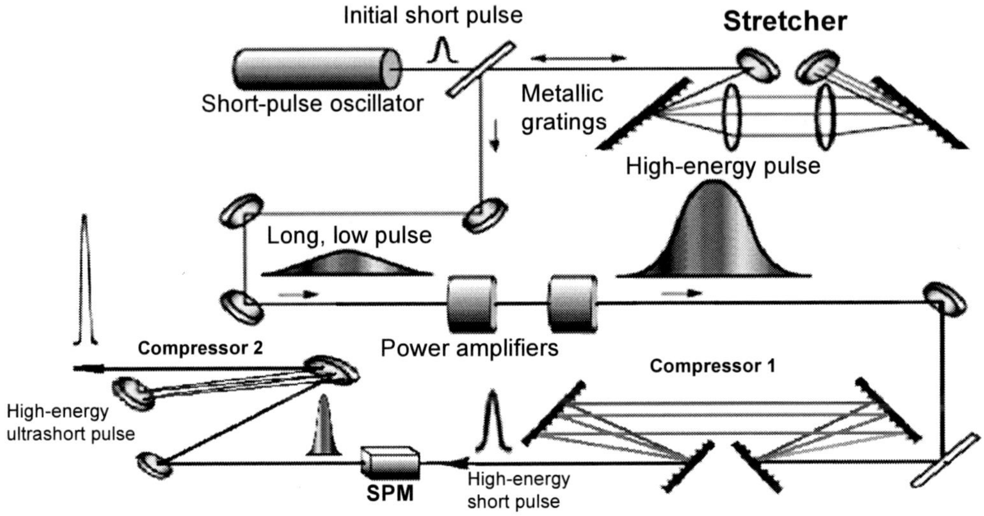

Figure 25. The secondary compression method.

We suppose that the next logical step could be suggested and consist of using super Gaussian beam profile because of higher uniformity [75]. If a laser beam has spatially uniform intensity, SPM can be used in near-field without spatial filtering and consequently eliminating constraints of the laser energy and transmission efficiency.

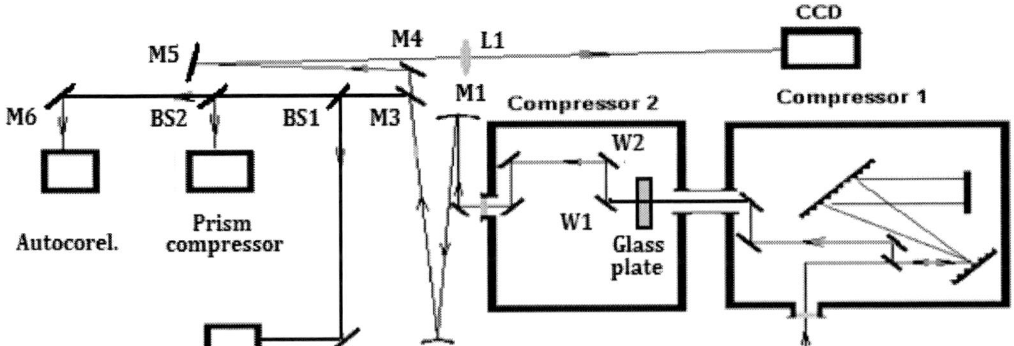

Figure 26. The experimental layout.

In a proof-of-principal experiment we demonstrated spectral broadening of a flat-top beam of the HERCULES laser [16] at 30TW power level to ~75 nm and compression of a fraction of this power from 30 to 14 fs by a prism-based compressor.

Figure 26 shows the experimental layout. The laser pulse with the energy about 1J, and flat top beam profile (Figure 27a) of 6 cm in diameter compressed by the standard grating compressor to 30 fs was directed at intensity ~1 TW/cm^2 on the bulk of fused silica plate located in the compressor chamber 2. Glass plates of 0.8, 2 and 2.5 cm thicknesses were used in our experiments. After SPM in the glass, the incident beam energy was attenuated ~100 times by two wedges W1, W2 in order to use a prism-based compressor and eliminate distortions during measurements. Then, the M1, M2 telescopes reduced the beam diameter to 1 cm and directed it into the prism compressor. Far field measurements (M4, M5, L1 and CCD camera) of the leak through the mirror M3 have shown no significant difference in the focal spot with or without SPM.

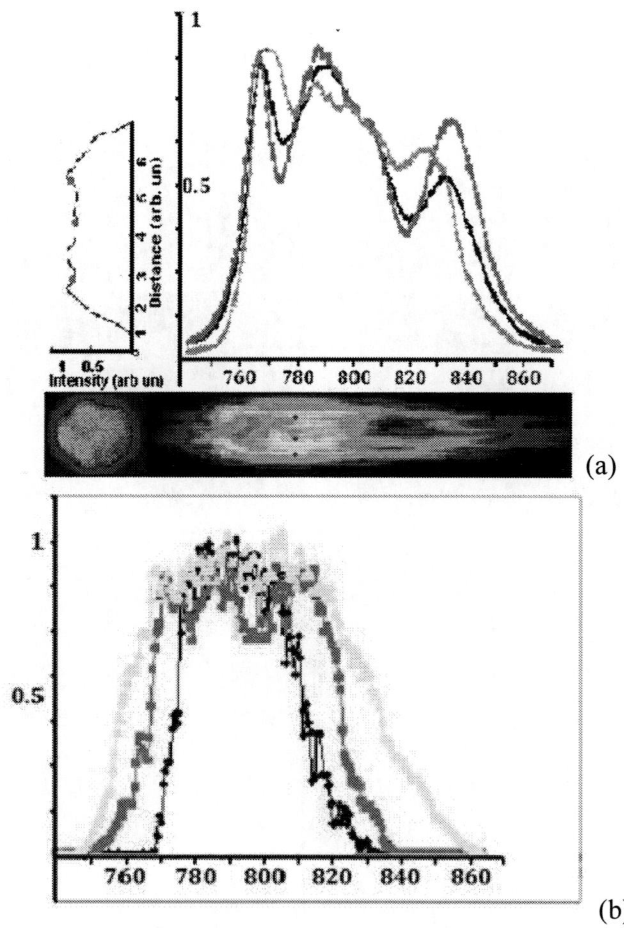

Figure 27. a-bottom picture: on the left-output beam of the HERCULES laser at 6J output energy, on the right spectrally resolved beam passed through 2cm glass plate, graphs above the spectrum show lineout of this spectrum taken at the spatial points of spectrally integrated spatial energy distribution shown as red dots on the graph in the upper left side ;b - spatially integrated pulse spectrums: blue curve (diamonds) represent the spectrum without the glass plate (no SPM), pink curve (squares) – with plate of the 0.8 cm thickness and yellow curve (triangles) – with glass plate of 2 cm thickness.

Experiments and Results

The spatially resolved spectrum of the pulse was measured and is shown on the picture in the bottom of Figure 27a. The lineouts of the spectrum at three spatial points in the beam profile (red spots on in-cut) are shown above the spatially resolved spectrum in figure 27a.

The graph of spatial energy distribution (in-cut) shows the deviation of the energy within 6% through most part of the beam aperture. It follows from Figure 27b that the initial FWHM of the space-integrated spectrum ~35nm was broadened to ~ 75nm with the 2 cm glass plate. The final compressor consists of two prisms (glass SF10) separated by ~ 50cm. The pulse duration is measured with a single shot autocorrelator. The autocorrelation picture and the trace lineout are shown in Figure 28a, b and c respectively.

Autocorrelation (a) and the solid blue curve in plot (c) show the initial nearly-transform limited pulse, with the duration from the FWHM autocorrelation trace of ~30 fs inferred assuming Gaussian pulse-shape. Autocorrelation (b) and the dashed pink curve in plot (c) shows the pulse after the SPM used with the 2.5 cm glass plate and recompression. The compressed pulse duration exhibits shortening by a factor of 2 leading to ~14 +/- 1 fs.

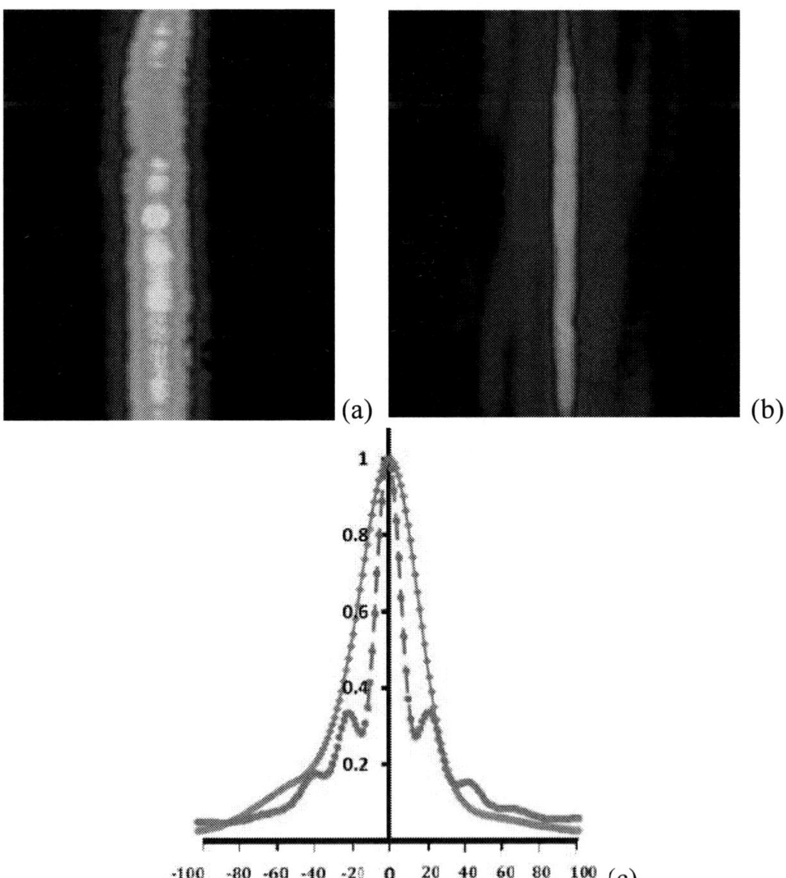

Figure 28. a- the initial transform limited pulse, b - compressed pulse after SPM, c – autocorrelation trace of the initial transform limited pulse (solid blue line) and compressed pulse after SPM (dashed pink line).

Discussion

During our experiments, we observed several complications, which prevent immediate implementation of this method into a real working system and require additional more precise investigation. These include the instability from shot to shot of the output pulse shape that manifested itself on the autocorrelation picture, the distribution of the significant part of the energy into wings and high level of the white light generation (up to 30% of the incident energy).

Nevertheless, we believe these are not crucial problems that prevent the application of this method due to the considerations below.

Obviously, the nature of experimentally observed instability of output pulse duration and its shape lies in inability to provide absolutely stable input pulse to the SPM process, but this original instability experiences significant growth during SPM and second compression. The wings oscillation of the output pulse could have a source in deviation from linearity of the chirp and Gaussian shape of the output spectra after SPM, however calculations demonstrate a significantly lower amplitude of this oscillations compared to the observed one. The right insert in Figure 29 shows Fourier transformation one of the measured spectra and its convolution into autocorrelation where the oscillation amplitude is more than six times lower than the observed one. Further estimations showed that a more probable source of the distortion of the output pulse shape and its instability is uncompensated third order dispersion (TOD) of prism compressor that was used as second stage of compression.

The following form of nonlinear Schrödinger equation governs a propagation of laser pulse through transparent dielectric media without absorption.

$$i\frac{\partial A}{\partial z} = D^2\frac{\partial^2 A}{\partial \tau^2} + iD^3\frac{\partial^3 A}{\partial \tau^3} - N|A|^2 A \tag{8}$$

Where $A(z, \tau)$ – slow varied amplitude, $\tau = (t - \beta_1 z)/T_0$, t and z – time and spatial coordinates of pulse distribution respectively $D^2 = \frac{|\beta_2|}{2T_0^2}$, $D^3 = \frac{|\beta_3|}{6T_0^3}$, $N = \frac{n_2\omega_0 I_0}{c}$, $I_0 = A_0^2$, T_0 – peak intensity and duration of the incident pulse, $\beta_1, \beta_2, \beta_3$ – group delay, group delay dispersion, third order dispersion, ω_0 – carried frequency, n_2- nonlinear index of refraction, and c –speed of light.

For pulse with the intensity of ~1 TW/cm^2, 30fs pulse duration and fused silica for bulk material we have coefficients of the right hand terms of the equation (8) - $D^2 = 0.04$, $D^3 = 0.004$, N = 2. This means the main role, in the case of SPM, plays the intensity depended nonlinearity. The first term of the right hand side of the equation influences onto the duration of the propagated pulse but not on its form. Further, the second term has a very low coefficient and can be neglected, and thus for our estimation we can ignore both of the first two terms. Resulting equation can be solved analytically. The solution for the Gaussian pulse can be done using the next formulas. $A(z, \tau) = A(0,\tau) \exp(i\,|A(0,\tau)|^2 zN)$, where $A(0,\tau) = \exp(-\tau^2/2T_0^2)$. Therefore, the Gaussian shape of the pulse remains Gaussian after propagation in the glass plate.

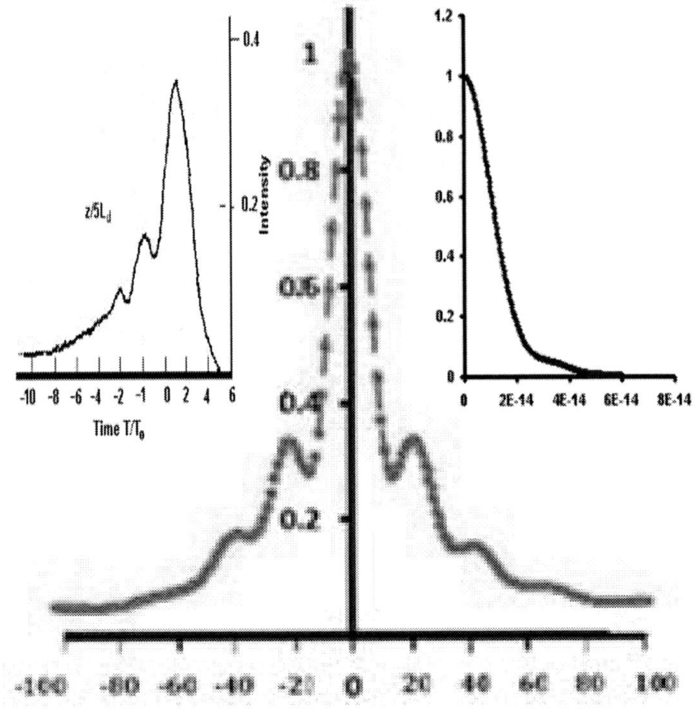

Figure 29. Autocorrelation trace of compressed pulse after SPM, left insert - numerical solution of nonlinear Schrödinger equation for media with parameters similar to prism compressor, right insert - Fourier transformation one of the measured spectra after SPM.

On the other hand, the used prism compressor possessed high dispersion properties, for example the TOD parameter equals β_3 = -14553 fs^3. In propagation equation (8) for equivalent media the dispersion terms have much larger coefficients: D^2 = -1.8, D^3= -1.1, N = 0.6 which is for the TOD term coefficient almost 300 times larger than the same parameter for fused silica, thus having a much stronger influence on the propagation process. Numerical solution under similar conditions shows that the shape of the output pulse is no longer Gaussian [76] and experiences high distortion that appears as a long oscillating tail (left insert in Figure 29), which can be seen on the autocorrelation trace as symmetrical wings (Figure 29).

In addition, TOD coefficient is inversely proportional to the third power of the pulse duration and the pulse duration consequently varies inversely with incident peak intensity. This is why even small fluctuation of the incident intensity leads to significant distortion of the shape of the compressed pulse.

A solution of these two complications could be in replacement of the prism compressor with chirped mirrors with compensated TOD. Using about 10 bounces on the chirped mirrors and the glass plate at Brewster angle promises compressor throughput above 90%.

In discussing efficient white light generation we would like to notice that it was also observed earlier, for instance in paper [77]. In our case, there is no signature of white light generation in the spectrum measurements (Figure 27) for two reasons. First of all, the white light power spectral density is at least four times lower than the main pulse because of the width of the spectrum. Second, the divergence of white light generation is much larger (12^0 in our experiments) than that of the laser beam, and so the three meter distance between the glass plate and the spectrometer (Figure 26) gives reduction of the relative intensity by more than 200 times. So overall, the intensity spectral density of the white light will be about three orders lower than the laser pulse.

Reducing white light generation could also be possible since the main processes involved in this generation are efficient only when phase matching conditions are fulfilled. In our case these conditions are intensity dependent [78], so reducing white light generation might be achievable by correctly choosing incident intensity and glass plate thickness. Evidence of such possibility can be found in the reference [74], where the SPM was done with a 3mm glass plate and incident intensity of 8 TW/cm^2, which resulted in a very low level of the white light generation.

As result, we suggest using super-Gaussian beam profile for CPA + SPM double compression method to avoid limitation of the incident energy and to increase transmission efficiency during recompression. We demonstrated more than doubling of the spectral pulse width via SPM, while maintaining a spatially uniform spectrum for a 30TW- laser pulse with super-Gaussian beam profile. In proof-of-principle experiment we demonstrated the possibility of pulse compression from 30 to 14 fs for 30TW- laser pulse. This scalable compression method if used with chirped mirrors may allow generation of ~10fs pulses at petawatt and exawatt power levels.

CONCLUSION AND FINAL REMARKS

The 50-year development of lasers, and the 27-year evolution of the CPA laser systems demonstrates incredible growth of the output power and intensity. The CPA-laser (Hercules [16]) demonstrated for the first time the power to 300 TW at a high repetition rate, as well as the highest intensity (~$2*10^{22}$ W/cm^2) and highest contrast (~ 10^{11}) [37]. The several important bottle-neck problems were solved during development of this laser system. Among these are:

- *The limitation for output energy*

  Our estimations reveal the TASE is a stronger restriction then transverse parasitic generation for increasing the aperture of the laser amplifiers. Nevertheless, the EDP-amplifiers, when operated under optimal conditions, are capable of suppressing the losses connected with transverse ASE and parasitic lasing, and increase the extracted energy by up to four times for four-pass amplifier, with a good efficiency, as compared to the conventional amplification with the same initial pump fluence. The existence of an optimal pump aperture for each given index-matched absorber and incident energy was demonstrated. CPA laser systems with EDP-amplifiers recently approached a record output power of 1.5 PW. With available index-matched liquid absorbers and large aperture Ti:sapphire crystals, it is possible to approach the kJ-level of extracted energy by exploiting EDP. With 60% compressor transmission efficiency and 10-15 fs pulse duration, 30 - 50 PW power and 10^{24}- 10^{25}W/cm^2 intensity could be reached from a single channel [50, 51].

- *The output pulse contrast cleaning.*

  For the laser pulses with such high intensity to be applicable for the laser-matter interaction experiments, a very high contrast is required. We have demonstrated that pulse cleaning based on XPW generation in two BaF_2 crystals yields a 10^{11} contrast ratio [37]. Dispersion compensation and direct amplification of the oscillator output are necessary for efficient XPW generation using only a microjoule energy. Two BaF_2 crystals placed in series increase the XPW conversion efficiency by a factor of 3 compared with a single crystal, yielding 3 orders of magnitude of contrast improvement. Further addition of the booster two-pass amplifier [16] and the double plasma-mirror after compressor [62] allowed achieving a contrast ratio of 10^{15} with output power of 300TW.

- *Pulse Front Tilt*

  Non-collinearity of the grating pairs could produce a pulse front tilt of the output laser pulse that can considerably increase the pulse duration at focus. We introduced a new, very simple, accurate and practical method for compressor alignment and demonstrate its advance for grating size adequate for compressors of Petawatt-scale lasers [65].

- *Wave front distortion.*

  The ultrahigh intensity is more effectively reachable by the tight focusing of the laser pulse, rather than by further power increase. However, tighter focusing readily affected by aberrations, so the compensation of the wave front aberrations during propagation through all systems and focusing optics is very desirable. We have demonstrated that by focusing a 300 TW laser beam to a spot size of 1.3μm a record peak intensity of $2 * 10^{22}$ W/cm^2 can be reached. To obtain such a tight focus with an excellent Strehl ratio of 0.9 we use an $f/1$ paraboloid and correct it by a 177-actuator deformable mirror. The direct wave-front correction is applicable only in a laser system that produces a stable shot-to-shot wave front. We have demonstrated a technique based on accurate wave-front measurement that can provide amplitude and phase information anywhere in the focal volume with high spatial resolution. We believe that the full characterization of a high-intensity beam will provide a new tool for accurately simulating experiments in this intensity regime. We believe that it will be possible to demonstrate an intensity of 10^{23} W/cm^2 following an upgrade of Hercules laser system.

- *Gain narrowing during amplification.*

  Pulse duration of high-power CPA lasers is strongly limited by gain-narrowing in the gain medium (to around 30 fs for 100 TW-PW lasers based on Ti:Sapphire). We suggest using super-Gaussian beam profile for CPA + SPM double compression method to avoid limitation of the incident energy and to increase transmission efficiency during recompression. We demonstrated more than doubling of the spectral pulse width via SPM, while maintaining a spatially uniform spectrum for a 30TW- laser pulse with super-Gaussian beam profile. In proof-of-principle experiment we demonstrated the possibility of pulse compression from 30 to 14 fs for 30TW- laser pulse. This scalable compression method, if used with chirped mirrors, may allow generation of ~10fs pulses at petawatt and exawatt power levels.

– *Next steps in the future of the high power lasers*

Concluding our discussion, we can now fix the current position of the highest laser power and intensity as 1.5PW [50, 51, 15] and 10^{22} -10^{23} W/cm^2 [28, 45, 16] respectively. Nevertheless, the existing technology allows to improve these parameters significantly. Applications of EDP-amplifiers as the final amplifier of the CPA laser systems with the Ti:Sapphire crystals as large as 20-30 cm diameter [54], and the recently developed liquid absorbers [26, 79] are able to help achieving an output energy of several kilojoules (Figure 15b). The capability to compress the pulse duration down to 10fs [75] carries the output power forward to the level of 100 PW from a single channel and to the intensity around 10^{25} W/cm^2. Multichannel laser system, using the same master-oscillator and coherent phase control could be able to sum the output power of all their channels – therefore only 10 channels will be enough to achieve Exawatt (10^{18} W) output power and tens of EW becomes reachable if a large-scale facility like NIF [12] is to be built. This can lead to an intensity as high as 10^{27} W/cm^2 and to the energy of accelerated ions around 10s of TeV, (Figure 2) if the laser is used as a laser-plasma accelerator.

The path to zettawatt level pulses (10^{21} W) was recently suggested in [80]. It was called C^3 for Cascaded Conversion Compression, where a compression method, weaving the three basic compression techniques, CPA, OPCPA and Plasma Compression by Backward Raman Amplification (BRA) in plasma, was proposed. The resulting intensity will approach the Schwinger value, thus opening up new possibilities in fundamental physics.

REFERENCES

[1] T.H.Maiman *Phys. Rev.* 123, 1145 (1961).

[2] W.Tiffany at all, *Appl. Phys. Letters,* 15, 91 (1969).

[3] A.Beaulie at all, *Appl. Phys. Letters,* 16, 504 (1970).

[4] K.Hohla at all, *Appl. Phys. Letters,* 16, 235 (1970).

[5] R. Yamamoto at all, *UCRL-CONF*-225230 (2006) .

[6] Stuart J. McNaught at all, *CLEO-* 2009, CThA1.pdf, (2009).

[7] John M. Slater *CLEO-2010, JThH1,* (2010).

[8] F.J. McClung, R.W. Hellwarth, *Proc. IRE 51*, 46 (1963).

[9] H.W. Mocker, R.J. *Collins Appl. Phys. Letters*, 7, 270 (1965).

[10] T. Tajima and J.M. Dawson. *Phys. Rev. Letters,*43, 267 (1979).

[11] *Nuckolls at all Nature*, 239, 129 (1972).

[12] https://lasers.llnl.gov/about/nif.

[13] http://www-lmj.cea.fr/index.html.

[14] D. Strickland and G. Mourou. *Optics Communication* 56, 3,219 (1985).

[15] J. H. Sung at all. *Optics Letters*,35 18, 3021 (2010); *Optics Express*, 20, 10807 (2012).

[16] V. Yanovsky at *all Optics Express* 16, 2109 (2008).

[17] Bell Sys. Tech. J., v. 39, pp. 754-808(1960).

[18] J. A. Giordmaine, at all *IEEE J. Quantum Electronics,* QE-4, 252 (1968).

[19] F. Gires and P. Tournom, *Compt. Rend. Acud. Sei.* (Paris), 258, 6112.

[20] M. A. Duguay and J. W. Hansen, *Appl. Phys. Letters,*.14, 14 (1969).
[21] E,B. Treacy. *IEEE J. of QE,* 5, 9, 454 (1969).
[22] O.E. Martinez at all. *J. Opt. Soc. Am.* A 1 (10), 1003 (1984).
[23] P. F. Moulton, *J. Opt. Soc. B*, 3, 125 (1986).
[24] *Optics Letters* 20, 389 (1995).
[25] *Applied Optics,* v.39,15, 2422 (2000) .
[26] F. Ple, M. Pittman, G. Jamelot, J.P. *Chambaret, Optics Letters* 32, 3, 238 (2007).
[27] http://www.clf.rl.ac.uk/New+Inisiatives/Upgrades+to+Existing+Facilities/.
[28] S.-W. Bahk, at all, *Opt. Lett.* 29, 2837 (2004).
[29] 3A. Zhidkov at all. *Phys. Rev. Lett.* 88, 185002 (2002).
[30] B. C. Stuart at all, *Phys. Rev.* B 53, 1749 (1996).
[31] M. D. Perry at all, *Opt. Lett.,* 24, 160 (1999).
[32] Gerstner, “Extreme Light”, *Nature,* 446, 16 (2007); http://www.eli-laser.eu/ .
[33] G.Pretzler at all, *Appl. Phys.* B 70, 1-9 (2000).
[34] F. Druon at all, *Opt. Lett.* 23, 1043 (1998).
[35] K.Yamakava, at all. *Optics Letters* 23, No. 7, 1 (1998).
[36] N.Ishii at all. *Optics Letters,* 30, 5, 567 (2005) .
[37] V. Chvykov at all *Opt. Lett.* 31, 1456 (2006).
[38] A. Jullien, *Opt. Lett.* 30, 920 (2005).
[39] S. Bank, at *IEEE J. Quantum. Electron.* 36, 268 (2000).
[40] V. Yanovsky at all *IEEE J. Sel. Top. Quantum. Electron*, 7, 539 (2001).
[41] V. Yanovsky at *CLEO-2003,* CME6 (2003).
[42] M. Aoyama at all *Opt. Lett.* 28, 1594 (2003).
[43] . F. G. Patterson at all *Opt. Lett.* 24, 963 (1999).
[44] Z. Sacks at all *Opt. Lett.* 26, 462 (2003).
[45] S.-W. Bahk at all *Appl. Phys.* B 80, 823 (2005).
[46] V. Chvykov at all *at Pross.* AAC2012, WG8, (2012).
[47] K.Ertel at all, *Optics Express* 16, 11, 8039 (2008).
[48] F.G. Peterson, at all *Optics Letters*, 24 14, 963 (1999).
[49] M.P.Kalachnikov at all *Laser Physics,* 12, 368 (2002).
[50] V. Chvykov at all OSA Technical Digest *CLEO 2003*, CWA34 (2003).
[51] V. Chvykov, K. Krushelnik *Optics Communications* v.285, 8, 2134 (2012).
[52] W. Koechner, *Solid State Laser Engineering, Springer,* Berlin, 1996.
[53] I.M.Frantz, J.S.Nodvik: *J. Appl. Phys*. 34 p. 2346 (1963).
[54] D. Joyce at all *Laser Focus World,* 46, 3, 37 (2010).
[55] J. Itatani at all *Opt. Commun.* 148, 70 (1998).
[56] V. Yanovsky at all *CLEO,* 39, CWK2. (2000).
[57] D. Homoelle at all *Opt. Lett.* 27, 1646 (2002).
[58] A. Jullien at all *Opt. Lett.* 29, 2184 (2004).
[59] M. P. Kalashnikov at all *Opt. Lett.* 30, 923 (2005).
[60] N. Minkovski at all *Opt. Soc. Am.* B 21, 1659 (2004).
[61] G. Doumy at all *Phys. Rev.* E 69, 026402 (2004).
[62] F. Dollar, Doctoral dissertation, *ProQuest Diss. And Thes.* (2012) .
[63] Qihua Zhu, at all *Facility Journal of Physics: Conference Series* 72, 012009 (2007).
[64] J.Qiao, at all *CLEO-2008, JThB5 (2008).*
[65] V. Chvykov at all *CLEO-2009, JWB4 (2009).*

[66] A. K. Sharma at all *Optics Express* , 14, 26, 13131 (2006).
[67] P. Varga and P. Török, *J. Opt. Soc. Am.* A 17, 2081 (2000).
[68] D. S. Montgomery at all *Phys. Rev. Lett.* 84, 678 (2000).
[69] P. Agostiniat all *IEEE J. Quantum Electron* QE-4, 667 (1968).
[70] B. Rau at all *Phys. Rev. Lett.* 78, 3310 (1997).
[71] N. M. Naumova at all *Phys. Rev. Lett.* 92, 063902 (2004).
[72] C. Rolland and P.B.Corkum, *J.OSA B,* 5, 641 (1988).
[73] M. Nisoli, at all *Appl. Phys.Lett.* 68, 2793 (1996).
[74] E. Mével, at all *J. Opt. Soc. Am.*B; 20 105 (2003).
[75] V. Chvykov at all *CLEO-2010, JThG4, (2010).*
[76] G.P.Agrawal, pp.98, *"Nonlinear Fiber Optics"* Academic Press, 4-th ed.
[77] A. Dharmadhikari, at all *Optics Express,* 12, 4, 695 (2004) .
[78] R.H. Stolen and J. E. Bjorkolm *IEEE J. of Qu. El.,* 18, 7, 1062.
[79] S. Laux, at all *Optics Letters,* 37, 1913, (2012).
[80] G. Mourou at all *Opt. Comm.,* 285, 720, (2012).

In: Laser Technology, Applications and Future Prospects ISBN: 978-1-62417-088-1
Editors: D. Ebersbach and F. Gottschalk

Chapter 2

LASER TECHNOLOGIES APPLIED FOR SEMICONDUCTOR HGCDTE- AND CHALCOGENIDE THIN FILM BASED STRUCTURES

H. Khlyap[1,*] and V. Laptev[2]
[1]University, Kaiserslautern, Germany
[2]Russian New University, Moscow, Russian Federation

ABSTRACT

CdHgTe-based semiconductors and structures as well as chalcogenide-based thin film active elements are of special interest for prospective applications in various fields of optoelectronics and information storage. Laser technologies made it possible to manufacture nanoscaled devices for far-infrared optoelectronics (CdHgTe and ZnCdHgTe solid solutions and thin films) and for optical information storage (Ag(Tl)Sb-containing thin films). The chapter reviews the latest results in fundamental and experimental research and reports unique data obtained by the authors. In particular, the growth technique and room-temperature electrical characteristics of ZhHgCdTe thin film based structures prepared by the pulse laser deposition and results from the investigations of laser technique-assisted structure engineering are presented. Applications of pulse laser deposition technology for producing long-life optical information storage media illustrate a wide perspective of manufacturing low-expensive and reliable active elements meeting needs of novel information storage.

1. INTRODUCTION

Laser technologies play one of the most important roles in the novel solid state nano-(opto)-electronics.

The pulse laser deposition method allows the fabrication of cheap and reliable nanoscaled devices. In this chapter we would like to draw the attention of the respectful reader to some

* E-mail: hkhlyap@yahoo.com

thermodynamical aspects in the field. We consider heat transfer modeling in metal-semiconductor structures and the general description of thermodynamic media containing phases with alternated-sign pressure. As the laser technologies are successfully applied for preparing high-quality semiconductor wafers used for designing hybrid heterostructures, "biological object-semiconductors", attention should be paid to the living cell as a thermodynamic object, which potentially can function as a nanoscaled device. The thermodynamic description of cytosol in a living cell provides data on how the biological structure functions as an energy-conversion device. The cytosol has chloroplasts, which converse solar radiation into energy. The cytosol is treated as homogeneous phase equilibrium with a special feature: the pressure of one phase is positive and the pressure of the other is negative. From this point of view the cytosol is neither solution nor gel (or sol as a whole) regardless of its components (water and dissolved substances). This is its unique capability for selecting, sorting and transporting reagents to the proper place of the living cell without a so-called "pipeline". To provide the basis for this statement, the theoretical investigation of equilibrium and stability conditions for the medium with sign-alternative pressure is carried out using the thermodynamic laws and the "Gibbs" equilibrium criterion.

2. Heat Transfer Modeling in Metal-Semiconductor Structures

First of all we consider how one can model heat transfer in common metal-semiconductor heterostructures.

We know the technological troubles arise under the homogeneous monocrystals $Zn_xHg_{1-x}Te$ whose growth are more considerable than that in case of $Cd_xHg_{1-x}Te$ monocrystals preparation [1]. This fact is bound with peculiarities of the diagram temperature-composition of the pseudobinary compound HgTe-ZnTe. However, new epitaxial techniques allow obtaining the layers of solid solution $Zn_xHg_{1-x}Te$ on CdTe or $Cd_yZn_{1-y}Te$.

Layers of $Zn_xHg_{1-x}Te$ with $0.7<x<1.0$ can be grown by the liquid-phase epitaxy method on ZnTe substrates, whereas the material of composition $x<0.3$ is manufactured on CdTe wafers due to necessity of the layer and substrate lattice parameters match. The best results of the lattice matching were observed using solid solutions $Zn_zCd_{1-z}Te$ with the proper composition.

We have prepared the layers of $Zn_{0.86}Hg_{0.14}Te$ compound on (111)-oriented ZnTe substrates by the method of epitaxy from the stoichiometric melts [1]. The start temperature of the crystallization was about 738^0C and had been lowered by 3^0 during 5 min. As a result, the epitaxial layer of the solid solution ZnHgTe of 50 μm thickness was obtained.

Different semiconductor structures (diodes, photodetectors, optoelectronic active elements, metal-semiconductor structures) are based on various materials in particular, A^2B^6 compounds (ZnTe, CdTe, etc.), which play an outstanding role in modern microelectronics [1]. Their operation under electric fields applied to the active element is ensured by metallic (or intermetallic) contacts formed by means of varied techniques (vacuum deposition and fusion are of special interest due to the simplicity and reliability of the methods). Nevertheless, even the best contacts provide a sufficient heat flow through the operation area of the semiconductor heterostructure, which can lead to the significant changes of some principal characteristics of the device, especially, the non-ideality of the current-voltage dependence determining the main processes of the charge transport [2].

This manuscript presents first results of the heating process studies performed for the structure In-ZnTe/In-$Zn_{0.86}Hg_{0.14}Te$.

Samples chosen for experimental measurements were cut off from the as-grown ring plates as rectangular boxes of electric area ~4.5 mm^2. Indium contacts were deposited as homogeneous pads on both left and right end surfaces of the sample.

Current-voltage (I-V) examinations, performed under room temperature, were demonstrated using the exponential character of the I-V-dependence (Figure 1) described by the equation (2.1):

$$I = I_s\left(\exp\left(\frac{eV}{n k_B T}\right) - 1\right), \tag{2.1}$$

where I_s is a saturation current (a complete function determined by the electronic parameters of the components of the structure), V is an applied voltage, e is an elemental charge, T is the experiment temperature, k_B is the Boltzmann constant and n is a non-ideality factor strongly dependent on the effects of the structure warming, and a presence of the quasi-neutral region localized at the ZnTe-ZnHgTe interface [3]:

$$n = \frac{e}{k_B T}\frac{dV}{d\ln I}. \tag{2.2}$$

Numerical analysis of the experimental data carried out according to the methods developed in [3] pointed out significant non-ideality of the I-V characteristic caused by different heat effects that arise due to charge carriers warming [4]. Figure 2 plots the scheme of the investigated structure using the following presentation.

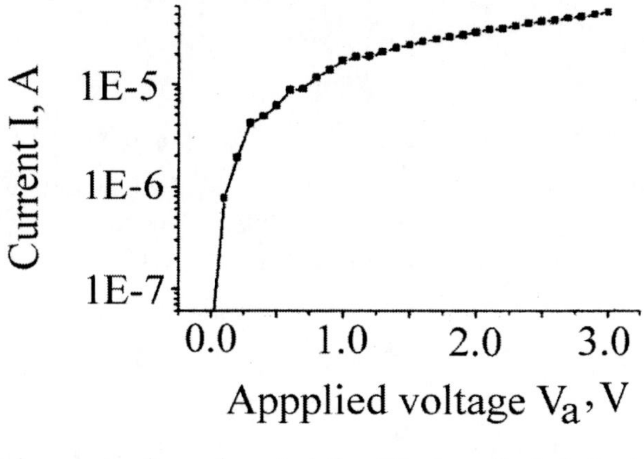

Figure 1. Experimental current-voltage characteristic of the investigated structure, $T_{thermostat}$ = 290 K.

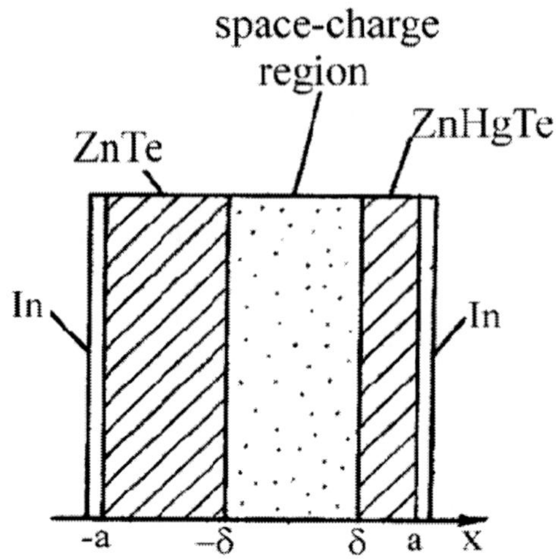

Figure 2. Schematic plot of the investigated structure: two metal-semiconductor structures In/ZnTe and In/ZnHgTe and the space-charge region.

The numerical value of n in our case changes from 1 under the applied electric field and increase to $5\cdot10^2$ V m^{-1} to 3 as the electric field increases (Figure 1). It is known [3], that such values of the non-ideality coefficient indicate the considerable changes in the mechanism of the charge carriers transport in the investigated structure from diffusion to tunneling through deep charge centers localized at the interface *substrate-film*.

These processes are caused by the fluctuations of the charge carrier concentration [2] due to thermal generation from the metal-semiconductor structure on both sides of the sample (see Figure 1). So, taking into account the heating processes of the two subsystems (electrons and photons) [4] one can perform the numerical simulation of the experimental results according to equation (2.3) assuming that the current (heating) contacts are placed at $x=\pm a$ under the same temperature (temperature of the thermostat):

$$\frac{eV}{nk_BT} = \ln\left\{\frac{j_s}{j_{s0}}\left[\exp\left(\frac{e\phi_b}{k_BT_{ph}} - \frac{e(\phi_b - V)}{k_BT_e}\right) - 1\right] + 1\right\} \tag{2.3}$$

where j_{s0} is the saturation current without field, ϕ_b is the potential barrier height (determined experimentally by the capacitance-voltage and current-voltage studies), T_e and T_{ph} are temperatures of electron and phonon subsystems, respectively. The theoretical I-V dependencies are described by the following equations:

$$I_1^{theor} = D_{tun} A_{el} j_{so} \exp\left(\frac{eV}{n_1 k_B T_{th}}\right), \tag{2.4}$$

$$I_2^{theor} = D_{tun} A_{el} j_s \exp\left(\frac{eV}{n_2 k_B T_{th}}\right), \tag{2.5}$$

where D_{tun} stands for the tunneling transparency of the potential barrier at interface ZnTe/ZnHgTe and is modulated by the thermal generation from the metal-semiconductor regions carriers, A_{el} is an electrical area of the sample (the area covered by the heat and charge carriers flows), T_{th} takes care of the thermostat temperature (determined by the conditions of the experiment). The case when temperatures of the electron and photon subsystems are equal to that of the thermostat under applied electric fields no more than $5 \cdot 10^2$ V m^{-1} seem to be of special interest for practical requirements. So, results of the numerical modeling are plotted in Figure 3. The data obtained from the experimental examinations and numerical modeling of electric field-induced heat effects in the metal-semiconductor structure are based on wide-gap A^2B^6 materials and demonstrate the sufficient influence of the thermally generated carriers on the current-voltage characteristic and the height of the potential barrier of the structure, in particular, on the value of the tunneling transparency coefficient D_{tun}, which varies from 10^{-8} to 10^{-4}. This fact can be explained by the thermal flip-over of the carriers' packets under the definite values of the applied electric field and appearance of the peculiar waves of the fluctuation of the charge carrier concentration.

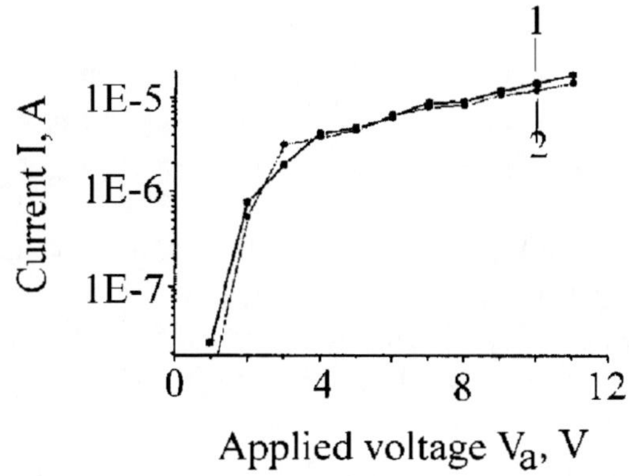

Figure 3. Theoretical I-V-dependence calculated according to the eq. (2.4), (2.5) in comparison with the experimental data.

3. Laser Technologies Prepare the Substrates for Future Hybrid Bio-Semiconductor Structures: The Living Cell as an Active Element

(Submitted by Dr. V. Laptev).

Cytosol in a living cell (intracellular fluid or cytoplasmic matrix, hyaloplasm matrix, aqueous cytoplasm) is a combination of the water-dissolved substances. It places itself in the cell between the plasma membrane, the nucleus and a vacuole; it is a medium keeping granular-like and whisker-like structures. Although water forms the large majority of the cytosol, its structure and properties within the cells is not well understood. The cytosol is always moving, it has about a half of the volume of the cell and one third of the cell mass. It supports the chemical transport, takes part in glycolysis, synthesis and destroying amino acids and in the protein (albumen) synthesis. The cytosol is a homogeneous solution and, at the same time, is a system containing the interacted parts [5-9]. As known, the Gibbs' phase rule explains freezing-out and segregation stability of seawater, but the unique stability of cytosol towards segregation and freezing-out has had no thermodynamic explanation until recently.

3.1. Know-How, Basic Thermodynamic Definitions

The paper reports the conditions of equilibrium and segregation stability of a medium consisting of a positive pressure phase and a negative pressure phrase. It is shown that this unusual state of the medium with the negative pressure and positive temperature is not metastable and coexists without segregation in equilibrium with a usual medium state where the pressure and temperature are positive.

The considered medium is matter as a homogeneous or a heterogeneous thermodynamic system. We suppose that the homogeneous medium can exist as a combination of two phases without segregation. This state is called homogeneous phase equilibrium. In this paper the properties of multiphase homogeneous medium are described.

3.1.1. Inertial Motion in U,S,V-Space

A full description of the thermodynamic state of a medium without chemical interactions is given by a relationship between the internal energy U, entropy S and volume V [5]. The mathematical procedure for negative pressure supposes using absolute values of internal energy U and entropy S.

The surface $\varphi(U,S,V) = 0$ corresponds to all equilibrium states of the medium. The isolated medium spontaneously reaches the thermodynamic equilibrium state at any point of this surface. The internal forces will bring the medium to equilibrium if the entropy will be maximized and the internal energy will be minimized [9 - 12].

This work supposes that the impulse of the internal forces arises in the medium during a time period of its motion to the equilibrium; this pulse in conserved under compensation of external forces or their absence. In this case instead of the rest state the inertial motion of the medium without work production is possible.

3.1.2. Zero and Negative Pressure

The inertial motion of the medium is described by the trajectory in the surface $\varphi(U, S, V) = 0$, which can intercept the regions of positive, negative or zero pressure in a general case. The effect of the negative pressure is known as a cavitation in fluids and a homogeneous tensile deformation of a solid. The mechanics are a hypothetical Chaplygin gas with a negative pressure. Cosmology operates with a category of negative pressure for a homogeneous isotropic medium. A cosmologic medium state with a zero pressure is called a "dust"-like one. However, we don't know an example of the medium or phase equilibrium with a sign-alternate pressure.

3.2. Cytosol as a Thermodynamic Medium

The cytosol is modeled as a homogeneous medium with a constant mass and without chemical interactions.

3.2.1. Cytosol in Equilibrium State

The cytosol in a living cell may be considered as a homogeneous medium in equilibrium. Actually, we can concern the relationship between the following parameters: the time of pressure stabilization, temperature and water density, sound velocity (v~1400 m/s), diffusion coefficient (D~0.2 m^2/s), thermal diffusivity (K~0.561 W/(m K)) and the linear dimension of the cell (L~20 μm). Then the relaxation time of the cell water relating to the pressure change L/v is $3x10^{-9}$s, the density relaxation time L^2/D is $2x10^{-11}$s and the temperature relaxation time is $L^2/K \sim 8x10^{-12}$s. The relaxation times in a water solution or cytosol are somewhat larger [9]. But in this case the time of metabolic processes is considerable exceeding the cytosol relaxation times to the equilibrium, i.e. these times are quasi-static.

3.2.2. Cytosol under Inertial Motion

The condition of the inertial motion of cytosol can be realized in a living cell. There are no principal contradictions to the statement that the cytosol in a living cell moves on the expense of internal resources. Really, the cytosol is surrounded by the plasma membrane and doesn't experience the deviations of the atmospheric pressure. Osmosis is produced by the cell pressure, which is compensated by pressure of the cell wall. There is neither hydrostatic pressure drop nor temperature gradient in the cytosol and the cell.

3.2.3. Cytosol as a System with Finite Energy

The fundamental concepts of energy and entropy for ecologic and biologic systems concern the living beings as open non-equilibrium thermodynamic systems permanently exchanging energy and matter with the environment [14-16]. We emphasize the fact that the energy resources and entropy of the living organism are finite. So the cell as the basic structural and functional unit of all known living organisms is considered a thermodynamic system with limited internal energy.

3.3. Theory

The first law of thermodynamics for systems with finite energy is unchanged as a law about entropy existing and increasing [12]. Pressure p and temperature T of the cytosol are determined by a plane tangent to the surface φ $(U,S,V)=0$ of our model medium. It is an algebraic surface $\varphi_1(U,S,V)=0$ of the first order in orthogonal coordinates, its slope to the surface $\varphi=0$ determines pressure p and temperature T by the equations $p= -(\partial U/\partial V)_S$ and $T=(\partial U/\partial S)_V$. The coordinates of the tangent points are given by the equation

$$U = TS - pV \tag{3.1}$$

after superposition of tangent surfaces at the origin of coordinates. When energy U is restricted, the sign of p and T alternates. The system of nuclear spins of a crystal with a total energy and entropy restricted by the spin orientation can serve as an example [12, 17]. Media with finite internal energy are known as unusual thermodynamic systems.

In algebra the general condition of thermodynamic equilibrium for usual and unusual phases is as follows: the virtual displacements of total entropy $\delta S^* = 0$ under constant U^* and fixed V [5]. For example, the isolated system from two phases 1 and 2 with ratios $S^*=S_1+S_2$ and $U^*=U_1+U_2$, $T\delta S=\delta U+p\delta V$ and $\delta U_1+\delta U_2=0$ is described by the equation

$$(1/T_1-1/T_2)\delta U_1 + (p_1/T_1)\delta V_1 + (p_2/T_2)\delta V_2 = 0 \quad (3.2)$$

with phase indices only.

3.3.1. Cytosol as an Usual Medium

It is known [10-12], that if $\delta V_1 = -\delta V_2$ the solution of (3.2) is

$$T_1 = T_2, \qquad p_1 = p_2,$$

In geometry these usual states 1 and 2 are identified by two tangent points of the plane to the surface $\varphi\,(U, S, V) = 0$ [9].

3.3.2. Cytosols an Unusual Medium

According to (3.1), the negative pressure is possible only under negative pressure. Nevertheless, the reverse speculation is not possible: one cannot state that the negative pressure exists under negative temperatures only. So, in the equilibrium thermodynamic system the alternating-sign pressure is possible under positive pressure. For example, the isolated system with $\delta V_1 = \delta V_2$ has a solution of (3.2):

$$T_1 = T_2, \; p_1 = -p_2. \quad (3.3)$$

Geometrically, it is presented by two tangent planes to the surface $\varphi=0$ with a total slope to the surface U=const is 180^0.

3.3.3. Medium Segregation Stability

We don't know any analysis of the stability of the alternating-sign pressure medium. Let us examine their stability related to the initial medium using the Gibbs' stability criterion [11,12]. The initial medium with constant mass is homogeneous and isotropic. Indexing its entropy $S_0=S_1+S_2$ we rewrite (3.1) as

$$U_0 - T_0 S_0 + p_0 V_0 = 0. \quad (3.4)$$

We have to find out which sign, =, > or < appears between the internal energy U_0 of the initial medium and the sum U_1+U_2 of the phase energies under the sign variation of the pressures p_1, p_2 and the pressure p_0 of the initial medium.

3.3.4. Segregation Stability of the Usual Medium

Let us consider the decay of the initial medium with formation of phases 1 and 2. If $V_1+V_2=V_0$ and $p_1=p_2$, expressions (3.1), (3.4) may be rewritten as

$$U_0-(p_{1,2}-p_0)\,V_0=U_1+U_2. \tag{3.5}$$

Then we examine the phase equilibrium stability relating the initial medium under variation of the pressure sign in (3.5) for three cases.

Scenario 1*i*. If $p_0=p_1=p_2$, instead of (3.5) we have

$$U_0=U_1+U_2.$$

Sign = here means that the initial medium and the phase can be in equilibrium under segregation. To visualize our description we denote the internal energy as $\bar{U}$ when the pressure is negative. Then the sign = in the expression below

$$\bar{U}_0=\bar{U}_1+\bar{U}_2$$

means that the initial medium and phases can be in equilibrium under negative pressure.

Scenario 2*i*. When $p_0=0$ and $p_1=p_2>0$, instead of (3.5) we have

$$U_0>U_1+U_2.$$

The sign ">" means that the initial medium is unstable relating to the equilibrium of new phases with a positive pressure under their segregation.

Scenario 3*i*. If $p_0=0$ and $p_1=p_2<0$, instead of (3.5) one has

$$U_0<\bar{U}_1+\bar{U}_2.$$

The sign < means that under segregation of 1 and 2 with negative pressure their equilibrium is unstable relating to the initial medium with a zero pressure.

3.3.5. Segregation Stability of an Unusual Medium

Now we ascribe the positive pressure to phase 1 and the negative pressure to phase 2.

In the geometry of the U, S, V – space it means the intercept of two planes with a unique tangency with the surface $\varphi=0$. The interception line is an isotherm of the plane V=const.

Then the equilibrium condition (3.3) is physically valid under absence of the phase segregation and $V_1=V_2=V_0$, where V_0 is a volume of the medium. In this case, according to (3.1), (3.3), (3.4),

$$U_0+p_0V_0=U_1+U_2. \tag{3.6}$$

Now we again investigate the phase equilibrium condition relating to the initial medium under variation of the pressure of initial medium p_0 in three cases below.

Scenario 1*j*. When $p_0 = 0$, instead of (3.6) we have

$$U_0 = U_1 + \bar{U}_2.$$

The sign = means that under $p_0=0$ the phases 1 and 2 may be in equilibrium with the initial medium without segregation.

Scenario 2*j*. If $p_0 > 0$, instead of (3.6) we have

$$U_0 < U_1 + \bar{U}_2.$$

The sign < means that the phases 1 and 2 are unstable relating to the initial medium with a positive pressure without segregation.

Scenario 3*j*. When $p_0 < 0$, instead of (3.6) we have

$$\bar{U}_0 > U_1 + \bar{U}_2.$$

The sign ">" means that the initial medium with a negative pressure is unstable relating to the equilibrium of new phases without segregation.

3.3.6. Compensation of Usual and Unusual Media

I-j scenarios have a different energy balance for the initial medium and new phases and suppose compensation. Assume that the initial medium spontaneously decays according to the scenario 3*j*. Corresponding to (3.5), the uncompensated energy is equal to the difference

$$\Delta U = \bar{U}_0 - (U_1 + \bar{U}_2) = -p_0 V_0.$$

When the initial medium spontaneously decays according to the scenario 2*i*, we have $\Delta U = U_0 - (U_1 + U_2) = p_{1,2} V_0$. The zero difference $\Delta U = 0$ corresponds to the scenario 1*i* or 1*j*. The uncompensated energy ΔU is spent on the medium mechanical reconstruction at constant p and T.

3.3.7. Homogeneous and Heterogeneous Phase Equilibrium

The Gibbs' conditions for equilibrium and stability of thermodynamic media with an alternating-sign pressure discovered in this work allow two rules to be built:

1. The equilibrium phases are stable to the segregation under inertial motion or in rest if they have pressure of opposite signs.
2. The equilibrium phases exist separately under the inertial motion or in rest if they have a pressure of the same sign.

A zero isobar from the scenarios *i-j* is a boundary curve between the homogeneous and heterogeneous phase equilibrium regions on the surface $\varphi\ (U, S, V) = 0$. Corresponding to (3.1), at each point of the zero isobar the product of temperature and $U/S \equiv 1/\alpha$ equals to unity.

According to the rules 1 and 2, the medium under inertial motion intercepts the zero isobar and excludes the phase segregation by itself when $T/\alpha \leq 1$.

3.4. Inertial Motion and Phase Rule

According to (3.1), the inertial motion of the medium takes place above the zero isobar if $U<TS$. For the zero isobar $U=TS$ and the spontaneous phase transition is possible corresponding to scenarios 1*i* or 1*j*. The inertial motion of the medium takes place below the zero isobars corresponding to scenario 3*j*.

Regarding (3.1) and Gibbs' phase rule, the one-component system in the heterogeneous equilibrium may be realized in no more than three phases. If the phases number reaches its maximum the inertial motion of the medium finishes due to absence of the thermodynamic degrees of freedom. According to scenarios 1*i* and Gibbs' phase rule, the inertial motion of the medium with positive pressure stops in triple point.

There is no evidence that the Gibbs' phase rule is not valid for homogeneous phase equilibrium. According to (3.3), the phases have different pressure and Gibbs had pointed out the constant mass ratio as a peculiarity of these phases [9]. It is also kept for an inertial motion. Regarding scenarios *j* and Gibbs' phase rule, the one-component system in the homogeneous equilibrium has one state with zero thermodynamic degrees of freedom. According to scenario 1*j*, three phases coexist under zero pressure only. So the zero isobar is a two phase boundary curve with other triple point for one-component system. The inertial motion finishes in this point.

3.5. Cytosol as a Homogeneous Phase Equilibrium

3.5.1. Cytosol and Zero Isobar

Going back to (3.1), the inertial motion of solution, gel or sol takes place above the zero isobar if $U<TS$. For the zero isobar $U=TS$ and the spontaneous phase transition it possibly corresponds to scenarios 1*i* or 1*j*. The new homogeneous or heterogeneous phase equilibrium is now neither gel nor sol or solution despite the fact that water and dissolved substances are its components.

The phase transition heterogeneous-homogeneous equilibrium is known in the cell physiology. The protein the coagulation of the albumen in the cytosol takes place at 42^0C. On the other hand, the cells function up to water freezing-out under cooling down to -89^0C in nature and down to -196^0C in the laboratory. This work supposes that the cytosol is not segregated and does not freeze out water because of its inertial motion along the zero isobar and downwards.

A thermodynamic surface $\varphi\ (U,\ S,\ V) = 0$ describes a thermodynamic equilibrium of water with a positive pressure and albumen mass with a negative pressure. For a negative pressure phase $\bar{U}>TS$ is valid. Then in the inertial motion along the zero isobar and downwards ratios $U_1<TS_1$ and $\bar{U}_2>TS_2$ are valid. The combination of signs < and > shows a unique ability of cytosol for a space separation, sorting and transportation of reagents to the proper place of the cell without a "pipeline."

3.5.2. Parameter α in a Biological Experiment

The condition of minimum energy at the constant entropy of the cytosol is a governor of inertial motion. The cytosol intercepts the zero isobar at the condition $T/\alpha = 1$. Then $\alpha = 1/(42+273.15) = 0.00317\ K^{-1}$ at 42^0C and $0.013\ K^{-1}$ at the temperature of liquid nitrogen. From this point of view the homeostasis, as a regulation of the living cell chemical activity, has to support the value of the parameter $1/\alpha$ of the cytosol so that condition $\alpha T \leq 1$ is to be valid. The negative pressure appears at scattering the medium particles and positive pressure leads to their compression. Optical study demonstrates that the dye makes a cell wall only for a living cell in the agar-agar medium and it penetrates into the cytosol of the dead cell only. One can suggest that the negative pressure of the living cell moves out the dye. Then the cell dies due to the intercept zero isobars by the cytosol and pressure switch from negative to positive according to the scenarios $1i$ or $2i$. Now the dye can enter the membrane bag.

The cytosol can be formed from the solution, gel or sol according to the scenarios $1i \rightarrow 1j$ in various ways. The experiment based on the biology of the living cell requires changing the parameter α of the initial solution in the way of helping to realize a transition through the zero isobars to prevent phase segregation.

In other words, biological processes between the components may be initialized in the laboratory if one can produce homogeneous phase equilibrium instead of solution, gel or sol, corresponding to (3.3).

3.5.3. Radiation in Biological Experiment

The zero isobar intercept leads to a change of the positive pressure with a negative one during the energy transition from the state $U<TS$ to the state $\bar{U}>TS$ at constant temperature. Without heat transfer the internal energy of the cytosol can be enlarged by means of metabolism or photosynthesis only.

The photosynthesis efficiency in the plant cell is found out in [18]. Remembering the nature photosynthesis realized by the living green leaf we should note that the thermodynamic model of the living structure proposed in this work can be a base for further fundamental research and practical studies of the possibility of designing principally novel systems based on biological objects.

4. The Equilibrium of the Phases with the Alternating-Sign Pressure

(*Submitted by Dr. V. Laptev*)

A complete description of the thermodynamic state of a medium with constant mass and without chemical interactions is given by a relationship between the internal energy U, entropy S and volume V [21]. They have the positive values. In thermodynamic S and V conjugate with temperature T and pressure p. There are thermodynamic systems with a positive and negative temperature [22]. No systems with an alternating-sign pressure are known. This paper describes the thermodynamic stability of the still unknown systems with opposite sign pressure.

4.1. Know-How, Basic Thermodynamic Definitions

This part of the chapter reports the conditions of equilibrium and segregation stability of a medium consisting of a positive pressure phase and a negative pressure one. It is shown that this unusual state of the medium with the negative pressure and positive temperature is not metastable and coexists without segregation in equilibrium with a usual medium state where the pressure and temperature are positive.

The considered medium is matter as a homogeneous or a heterogeneous thermodynamic system. We suppose that the homogeneous medium can exist as a combination of two phases without segregation. This state is called homogeneous phase equilibrium. In this paper the properties of a multiphase homogeneous medium are described.

An initial matter is considered as a homogeneous medium with a constant mass and without chemical interactions.

4.1.1. U, S, V- Space

*The surface φ (*U, S, V*) = 0 corresponds to the all equilibrium states of the medium. It has a tangency with* a plane which is an algebraic surface φ_1 (*U*, *S*, *V*) = 0 of the first order in orthogonal coordinates. The coordinates of the tangent points are given by the equation

$$U = TS - pV \tag{4.1}$$

after superposition of tangent surfaces at the origin of coordinates. The mathematical procedure supposes using absolute values of the internal energy *U* and entropy *S*. They are positive.

Pressure p and temperature T of the medium are determined by a tangent plane to the surface φ (*U*, *S*, *V*) = 0. It is an algebraic surface φ_1 (*U*, *S*, *V*) = 0 of the first order in orthogonal coordinates, its slope to the surface φ=0 determines pressure p and temperature T by the equations

$$p = -(\partial U/\partial V)_S, \tag{4.2}$$

$$T = (\partial U/\partial S)_V. \tag{4.3}$$

The index of the derivative points out a value, which is supposed to be constant under differentiation [21, 23 - 25].

4.1.2. Usual or Unusual System

The signs of the values *U*, *S*, and *V* are constant. If the internal energy *U* monotonically und unrestrictedly increases with entropy *S* and volume *V*, this thermodynamic system is called a usual thermodynamic system, temperature and pressure are considered as positive values [25].

According to (4.1) - (4.3), *p* or *T* has an alternating-sign if the internal energy *U* of a system has extremum. Media with finite internal energy are known as unusual thermodynamic systems [25]. The nuclear spin energy of a solid can be such an example because it has upper limes. The temperature of such spin system is an alternating-sign [22].

We consider the wanted medium as an unusual system because the internal energy limit is the primary reason for its alternating-sign pressure. The properties of a system with finite internal energy are specific.

4.1.3. Alternating-Sign Pressure and Temperature

The spin temperature of the crystal is alternating-sign; the crystal temperature and pressure are positive only [22]. Unfortunately, the spin pressure sign is not found in those works. It can be found with help of the (4.1), (4.2). Obviously its right half is positive at the negative temperature if the pressure is negative. Therefore spin energy can be a reason of the even stretch of solid, which is compensated by electromagnetic attraction of spin carries.

The effect of the negative pressure is known as a cavitation in fluids and a homogeneous tensile deformation of a solid. The mechanics are a hypothetical Chaplygin gas with a negative pressure. Cosmology operates with a category of negative pressure for a homogeneous isotropic medium. A cosmologic medium state with zero pressure is called a "dust"-like one. However, we don't know an example of the medium or phase equilibrium with a sign-alternating pressure.

4.2. Theory

The first law of thermodynamics for systems with finite energy is unchanged as a law about entropy existing and increasing [25]. We pay particular attention to the fact that from (4.1) it follows that negative T is possible only at negative pressure p. It should be clear that this statement does not allow making a so-called reversible statement: it is not correct to think that negative pressures are possible only at negative temperatures. In this work the alternating-sign pressure at the positive temperature is studied.

4.2.1. Gibbs' Criterion

The surface $\varphi\ (U, S, V) = 0$ corresponds to the all equilibrium states of the medium. In algebra the general condition of thermodynamic equilibrium for usual and unusual phases is as follows: the virtual displacements of total entropy $\delta S^* = 0$ under constant U^* and fixed V [21, 23-25]. For example, the isolated system from two phases 1 and 2 with ratios $S^*=S_1+S_2$ and $U^*=U_1+U_2$, $T\delta S=\delta U+p\delta V$ and $\delta U_1+\delta U_2=0$ is described by the equation

$$(1/T_1-1/T_2)\delta U_1 + (p_1/T_1)\delta V_1+(p_2/T_2)\delta V_2 = 0 \tag{4.4}$$

with phase indices only. For any values of variations δU_1 or δU_2 we find preliminary equilibrium conditions for the usual and unusual media:

$$1/T_1-1/T_2 = 0.$$

If $\delta V_1 = \pm\ \delta V_2$ from (4.4) we have

$$(p_1/T_1 \pm p_2/T_2)\delta V_{1,2} = 0$$

for any values of variations δV_1 or δV_2.

4.2.2. Gibbs' Equilibrium Condition for Usual Medium

It is known [23-25], that if $\delta V_1 = -\delta V_2$ the solution of (4.4) is

$$T_1 = T_2, \qquad p_1 = p_2,$$

In geometry these usual states 1 and 2 are identified by two tangent points of the plane to the surface φ $(U, S, V) = 0$ [21]. Phase temperature and pressure change not the sign; such medium is a usual thermodynamic system.

4.2.3. Gibbs' Equilibrium Condition for Unusual Medium

According to (4.1), negative pressure is possible only under negative pressure. Nevertheless, the reverse speculation is not possible: one cannot state that negative pressure exists under negative temperatures only. So, in the equilibrium thermodynamic system the sign-alternative pressure is possible under positive temperatures. For example, here we have found that the isolated system with $\delta V_1 = \delta V_2$ has a solution of (4.4):

$$T_1 = T_2, \ p_1 = -p_2. \qquad (4.5)$$

Geometrically it is presented by two tangent planes to the surface $\varphi=0$ with a total slope to the surface $U = \text{const}$ is 180^0. Phase temperature and pressure change the sign; such medium is an unusual thermodynamic system.

4.2.4. Medium Segregation Stability

We don't know any analysis of the stability of the sign-alternative pressure medium. Let us examine their stability related to the initial medium of using the Gibbs' stability criterion [23-25]. The initial medium with constant mass is homogeneous and isotropic. Indexing its entropy $S_0 = S_1 + S_2$ we rewrite (4.1) as

$$U_0 - T_0 S_0 + p_0 V_0 = 0. \qquad (4.6)$$

We have to find out which of signs =, > or < appears between the internal energy U_0 of the initial medium and the sum U_1+U_2 of the phase energies under the sign variation of the pressures p_1, p_2 and the pressure p_0 of the initial medium.

4.2.5. Segregation Stability of the Usual Medium

Let us consider the decay of the initial medium with formation of phases 1 and 2. If $V_1 + V_2 = V_0$ and $p_1 = p_2$, expressions (4.1), (4.6) may be rewritten as

$$U_0 - (p_{1,2} - p_0)\, V_0 = U_1 + U_2. \qquad (4.7)$$

Then we examine the phase equilibrium stability relating the initial medium under variation of the pressure sign in (4.7) for three cases. The signs of p_1 and p_2 are identical.

Scenario 1*i*.If $p_0 = p_1 = p_2$, instead of (4.7) we have
$U_0 = U_1 + U_2$.

Sign = here means that the initial medium and the phase can be in equilibrium under segregation. To visualize our description we denote the internal energy of $\bar{U}$ when the pressure is negative. Then the sign = in the expression

$$\bar{U}_0 = \bar{U}_1 + \bar{U}_2$$

means that the initial medium and phases can be in equilibrium under negative pressure.

Scenario 2*i*.If $p_0 = 0$ and $p_1 = p_2 > 0$, instead of (4.7) we have

$$U_0 > U_1 + U_2.$$

The sign > means that the initial medium is unstable relating to the equilibrium of new phases with a positive pressure under their segregation.

Scenario 3*i*. If $p_0 = 0$ and $p_1 = p_2 < 0$, instead of (4.5) one has

$$U_0 < \bar{U}_1 + \bar{U}_2.$$

The sign < means that under segregation of phases 1 and 2 with negative pressure their equilibrium is unstable relating to the initial medium with a zero pressure.

The pressure p_1 and p_2 in scenario 1*i* - 3*i* have identical signs. Then the phases 1 and 2 are usual media.

4.2.6. Segregation Stability of an Unusual Medium

Now we ascribe the positive pressure to the phase 1 and the negative pressure to the phase 2. In the geometry of the U, S, V – space it means the intercept of two planes with a unique tangency with the surface $\varphi=0$. The interception line is an isotherm of the plane V=const. Then the equilibrium condition (4.5) is physically valid under absence of the phase segregation and $V_1=V_2=V_0$, where V_0 is a volume of the medium. In this case, according to (4.1), (4.5), (4.6),

$$U_0 + p_0 V_0 = U_1 + U_2. \tag{4.8}$$

Now we again investigate the phase equilibrium condition relating to the initial medium under variation of the pressure of initial medium p_0 in three cases below.

Scenario 1*j*. When $p_0 = 0$, instead of (4.8) we have

$$U_0 = U_1 + \bar{U}_2.$$

The sign = means that under $p_0 = 0$ the phases 1 and 2 may be in equilibrium with the initial medium without segregation.

Scenario 2*j*. If $p_0 > 0$, instead of (4.8) we have

$$U_0 < U_1 + \bar{U}_2.$$

The sign < means that the phases 1 and 2 are unstable relating to the initial medium with a positive pressure without segregation.

Scenario 3*j*. When $p_0 < 0$, instead of (4.8) we have

$$\bar{U}_0 > U_1 + \bar{U}_2.$$

The sign > means that the initial medium with a negative pressure is unstable relating to the equilibrium of new phases without segregation.

The pressure p_1 and p_2 in scenarios *j* have opposite signs. Then phases 1 and 2 combine unusual media.

4.2.7. Compensation of Usual and Unusual Media

I-j scenarios have a different energy balance for the initial medium and new phases and suppose compensation. Assume that the initial medium spontaneously decays according to the scenario 3*j*. Corresponding to (4.5), the uncompensated energy is equal to the difference

$$\Delta U = \bar{U}_0 - (U_1 + \bar{U}_2) = -p_0 V_0 > 0.$$

When the initial medium spontaneously decays according to the scenario 2*i*, we have

$$\Delta U = U_0 - (U_1 + U_2) = p_{1,2} V_0 > 0.$$

The uncompensated energy ΔU can be dissipated in space. The zero difference $\Delta U=0$ corresponds to the scenario 1*i* or 1*j*.

4.2.8. Inertial Motion

The surface $\varphi(U, S, V) = 0$ corresponds to all the equilibrium states of the medium. The isolated medium spontaneously moves to the thermodynamic equilibrium state. It reaches this at any point of this surface. The internal forces will bring the medium to equilibrium if the entropy will be maximized and the internal energy will be minimized [23 - 25].

This work supposes that the impulse of the internal forces arises in the medium during a time period of its motion to the equilibrium; this pulse in conserved under compensation of external forces or their absence. In this case, instead of the rest state, the inertial motion of the medium without work production is possible. The inertial motion of the medium is described by the trajectory in the surface $\varphi(U, S, V) = 0$ which can intercept the regions of positive, negative or zero pressure in a general case.

4.2.9. Homogeneous and Heterogeneous Phase Equilibrium

The Gibbs' conditions for equilibrium and stability of thermodynamic media with a sign-alternate pressure discovered in this work allow for two rules:

1. The equilibrium phases are stable to the segregation under inertial motion or in rest if they have pressure of opposite signs.
2. The equilibrium phases exist separately under the inertial motion or in rest if they have a pressure of the same sign.

The first rule corresponds to the homogeneous phase equilibrium or unusual medium. The second rule corresponds to the heterogeneous phase equilibrium or usual medium.

4.3. Zero Pressure and Phase Equilibrium

Zero pressure does not mean zero temperature. The temperature corresponding to zero pressure, can be found with function

$$\omega \equiv p/u = \alpha T - 1, \tag{4.9}$$

that is obtained by substitutions $u \equiv U/V$, $s \equiv S/V$ and $\alpha \equiv s/u$ into (4.1). It has a useful property; that for any positive number α we can find a zero value of ω and positive temperature $T_\alpha = 1/\alpha$. Consequently the value ω at $p = 0$ and $T = 1/\alpha$ divides the surface φ $(U, S, V) = 0$ into two regions: one with positive pressure at $T > T_\alpha$ and one with negative pressure at $T < T_\alpha$.

The isolated medium moves with time into the thermodynamic equilibrium state and reaches it at the temperature T_0 and positive pressure. The medium cannot leave the rest state by itself. Under unusual conditions of the isolation, when external forces are absent, the primary medium can save the beginning mechanical impulse of the quasi-static expansion. Then the equilibrium medium will be cooled adiabatically, expanding by the inertion. In our case the medium is isolated, U_0 and S_0 are not variable, α=const and the function (4.9) is linear.

The beginning state of inertial motion of the primary medium is presented by point O, on Figure 4. In the geometrical interpretation of the medium inertial motion, we use the fact that the identical parameter $\omega = -1$ exists for all functions $\omega = \omega(T)$ at zero temperature. According to the equation (4.9), the primary medium has the phase stability under expansion along curve ***oa*** under scenario 1*i* until the temperature T_a.

At point ***a***, the medium decay on the true new phases is possible due to the scenario 1*i* or 2*i* or 1*j*. The scenarios 1*i* and 2*i* describe the usual medium; the unusual medium appears under scenario 1*j*.

Let's realize the scenario 1*j*. When the primary medium decay is finished, the true new phases have opposite pressure signs at temperature T_a in the equilibrium states 1 and 2. If $p_1 = -p_2$, then $\omega_1 = -\omega_2$ at the elementary condition $u_1 = u_2$. Then segment ***b***1 is part of function

$$\omega_1 = \alpha_1 T - 1, \tag{4.10}$$

then segment *b*2 is part of function

$$\omega_2 = -\alpha_1 T + 1. \tag{4.11}$$

These segments are the coexisting curves for phase 1 and 2. It is obvious that the equilibrium of phase 1 and 2 c with alternating-sign pressure is possible between the point *b* and *a* or in the temperature interval $\{T_b, T_a\}$.

Point *b* phase 1 and 2 c with alternating-sign pressure achieves the zero pressure and at the temperature $T_b < T_a$ they spontaneously decay on phases 3 and 4 under scenario 1*j*. The equilibrium of phase 3 and 4 with alternating-sign pressure is possible in the temperature interval $\{T_c, T_b\}$.

At the point *c* at $T=T_c$ distinction of properties of phase 3 and 4 disappears and a zero pressure medium is generated which spontaneously decays on phases 5, 6 with alternating-sign pressure.

Further, at $T=T_d$ and $\omega=0$ spontaneous decay of zero pressure medium on phase 7 and 8 with alternating-sign pressure is possible, and so on. This process can be called a fractal-like multiphase recombination or the process by which the equilibrium phases with alternating-sign pressure can achieve zero pressure again.

It is important that radiation cannot be completely transformed into condensate. The work must be done. However, the work is not a product of the recombination cycle *a1b2a* or the other one.

The medium can begin the inertial motion at point *2* on Figure 4. In this case the same fractal-like multiphase recombination takes place in further according to scenario 3*j* and the scheme, presented above, with the non-zero compensation.

4.4. Zero Isobar and Phase Equilibrium

According to Figure 1 and the conditions (4.5), for the homogeneous phase equilibrium the equalities

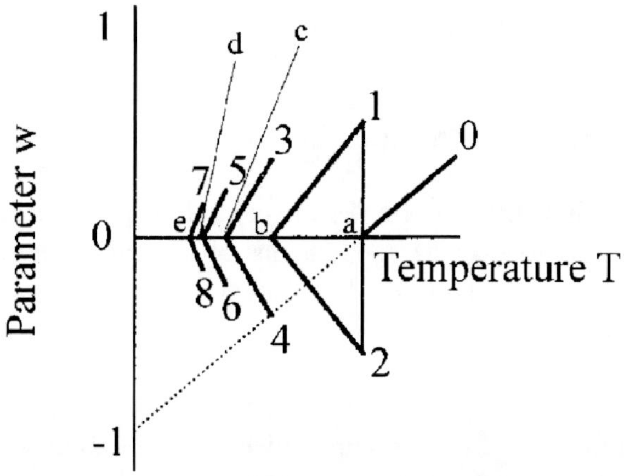

Figure 4. Illustration of function (4.9) and the inertial medium motion with and without alternating-sign pressure. Curve *ao* is the beginning stage of the usual medium motion. Points *a*, *b*, *c*, *d* and *e* are zero pressure of the states. Isotherms *1-2*, *3-4*, *5-6* and *7-8* represent the decay of zero pressure medium on alternating-sign pressure phases.

$\omega_1 + \omega_2 = 0$ at the $u_1=u_2$,

$\omega_3 + \omega_4 = 0$ at the $u_3=u_4$,

$\omega_5 + \omega_6 = 0$ at the $u_5=u_6$,

$\omega_7 + \omega_8 = 0$ at the $u_7=u_8$

are valid and the curve $\omega = 0$ is the zero isobar too. The curves on Figure 1 are the inertial motion trajectories of the phases 1 and 2 in the surface φ (U, S, V) = 0. In a general case, $p_1 = -p_2$ at the conditions $\omega_1 \neq -\omega_2$ and $u_1 \neq u_2$. Then the inertial motion trajectories of the phases 1 and 2 won't be straight lines in Figure 4. The role played by the zero isobar under homogeneous phase equilibrium is not changed.

A zero isobar from the scenarios *i-j* is a boundary curve between the homogeneous and heterogeneous phase equilibrium regions on the surface φ (U, S, V) = 0. So, according to the Gibbs' stability criterion, any medium with zero pressure coexists with zero pressure phases (scenario 1*i*). It is unstable relative to new phases with positive pressure (scenario 2*i*) and is stable relative to the new phases with negative pressure (scenario 3*i*). The new phases and initial medium are segregated.

Without segregation, the zero pressure mediums coexist with phases that have opposite pressure signs (scenario 1*j*). This results in the zero pressure curve dividing the surface φ(U, S, V)=0 into two parts which stand for the homogeneous states of the primary medium with positive pressure (scenario 2*j*) and other homogeneous states of secondary medium consisting of the new phases with alternating sign pressure which are not the primary medium (scenario 3*j*). If the first part of the surface is given, the other part can be calculated. Corresponding to (4.1), at the each point of the zero isobar the product of temperature and $U/S \equiv 1/\alpha$ equals to unity. According to rules 1 and 2, the medium under inertial motion intercepts the zero isobar and excludes the phase segregation by itself when $T/\alpha \leq 1$.

4.5. Inertial Motion and Phase Rule

According to (4.1), the inertial motion of the medium takes place above the zero isobar if $U<TS$. The zero isobar $U=TS$ and the spontaneous phase transition possibly corresponds to scenarios 1*i* or 1*j*. The inertial motion of the medium takes place below the zero isobar corresponding to scenario 3*j*.

Regarding (4.1) and Gibbs' phase rule, the one-component system in the heterogeneous equilibrium may be realized from no more than three phases. If the phases number reaches its maximum the inertial motion of the medium finishes due to absence of the thermodynamic degrees of freedom. According to scenarios 1*i* and Gibbs' phase rule, the inertial motion of the medium with positive pressure stops in triple point.

There is no evidence that the Gibbs' phase rule is not valid for homogeneous phase equilibrium. According to (4.3), the phases have different pressure and Gibbs pointed out the constant mass ratio as a peculiarity of these phases [21]. It is also kept for an inertial motion. Regarding scenarios *j* and Gibbs' phase rule, the one-component system in the homogeneous equilibrium has one state with zero thermodynamic degrees of freedom. According to scenario 1*j*, three phases coexist under zero pressure only. So, a zero isobar is a two-phase boundary curve with other triple point for a one-component system. The inertial motion finishes at this point.

4.6. Thermal Radiation and Condensate as Medium Claimers on Homogeneous Phase Equilibrium

4.6.1. Thermal Radiation as a Perl of Thermodynamics

Thermodynamics of the thermal radiation is a cradle of the quantum theory. Thermal radiation is a unique thermodynamic system while the expression

$$\mathrm{d}U = T\mathrm{d}S - p\mathrm{d}V$$

for internal energy U, entropy S, and volume V holds the properties of the fundamental equation of thermodynamics regardless of the variation of the photon number [23, 25]. The differential expression

$$\mathrm{d}p/\mathrm{d}T = S/V$$

for pressure p and temperature T is valid for a one-component system under phase equilibrium if the pressure does not depend on the volume V [24,25]. Thermal radiation satisfies these conditions but shows no phase equilibrium. However, successful attempts of finding thermal radiation condensate in any form are unknown too. This work aims to support enthusiasm of experimental physicists and reports for the first time the phenomenological study of the thermodynamic medium consisting of thermal radiation and its condensate.

4.6.2. The Hypothesis about Condensate of Radiation

It is known [23, 25], that evolution of radiation or its inertial motion is impossible without participating matter and it realizes this with absorption, emission and scattering of the beams as well as with the gravitational interaction. Transfer of radiation and electron plasma to the equilibrium state is described by the kinetic equation. Some of its solutions are treated as an effect of accumulation in a low-frequency spectrum of radiation, as Bose-condensation or a non-degenerated state of radiation [6-10]. A known hypothesis about Bose-condensation of relic radiation and condensate evaporation has a condition: the rest mass of photon is thought to be non-zero [31]. Nevertheless, experiments show that photons have no rest mass [32].

4.6.3. Thermal Radiation Stability

The stability determinant [25] of the equilibrium radiation

$$D = (\partial^2 U/\partial^2 S)_V(\partial^2 U/\partial^2 V)_S - \partial^2 U/(\partial S/\partial V)$$

is equal to zero because

$$U(S, V) = \sigma V(3S/4\sigma V)^{4/3}.$$

While the zero determinants are related to the limit of stability, there are no thermodynamic restrictions for phase equilibrium of thermal radiation [24].

4.6.4. Thermal Radiation Condensate as Usual Medium

Radiation, matter and condensate may form a total thermal equilibrium. According to the transitivity principle of thermodynamic equilibrium [23, 25], participating condensate does not destroy the equilibrium between radiation and matter. Suppose that matter is a thermostat for the medium consisting of radiation and condensate. A general condition of thermodynamic equilibrium is an equality to zero of virtual entropy changes δS or virtual changes of the internal energy δU for media [24, 25]. Using indices for describing its properties, we write

$$S=S_{\text{rad}}+S_{\text{cond}},$$

$$U=U_{\text{rad}}+U_{\text{cond}}.$$

The equilibrium conditions

$$\delta S_{\text{rad}}+\delta S_{\text{cond}}=0,$$

$$\delta U_{\text{rad}}+\delta U_{\text{cond}}=0$$

will be completed by the expression

$$T\delta S=\delta U+p\delta V,$$

and then we get an equation

$$(1/T_{\text{cond}}-1/T_{\text{rad}})\delta U_{\text{cond}}+(p_{\text{cond}}/T_{\text{cond}})\delta V_{\text{cond}}+(p_{\text{rad}}/T_{\text{rad}})\delta V_{\text{rad}} = 0. \tag{4.12}$$

If $V_{\text{rad}}+V_{\text{cond}}=V=\text{const}$ and $\delta V_{\text{rad}}= -\delta V_{\text{cond}}$, then for any values of variations δU_{cond} and δV_{cond} we find

$$T_{\text{rad}}=T_{\text{cond}}=T,$$

$$p_{\text{rad}}=p_{\text{cond}}=p$$

as the equilibrium conditions. According to scenario $1i$, condensate and radiation are segregated.

4.6.5. Thermal Radiation Condensate as Unusual Medium

When $V_{\text{rad}}=V_{\text{cond}}=V$ and $\delta V_{\text{rad}}=\delta V_{\text{cond}}$, the equilibrium conditions

$$T_{\text{rad}} = T_{\text{cond}},\ p_{\text{rad}} = -p_{\text{cond}} \tag{4.13}$$

are satisfied. According to scenarios j, condensate and radiation are not segregated. Condensate is absolutely transparent for radiation; the radiation is integrated into the condensate. The positive and negative phase pressure means that the energy of radiation and condensate has a limit. Then thermal radiation and condensate together are an unusual homogeneous medium.

4.6.6. The Entropy of Thermal Radiation and Condensate in Equilibrium

Let's consider the evolution of the condensate being in equilibrium with radiation. Once the medium is appeared, this medium consisting of the equilibrium condensate and radiation can continue the inertial adiabatic extension due to the assumed absence of external forces.

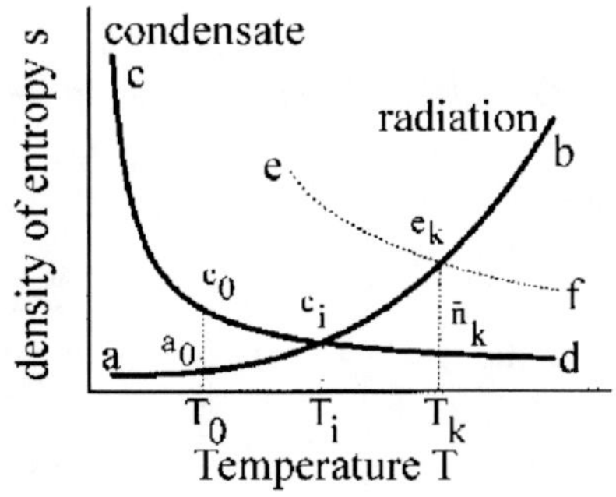

Figure 5. plots the density of entropy for radiation (curve *ab*) and for a condensate (curves *cd* and *ef*) are schematically plotted. The positive pressure of the radiation and negative pressure of condensate are equal by absolute value at points c_i and e_k at the interceptions of these curves.

Figure 5 plots a curve of radiation extension as a cubic parabola

$$s_{rad}=4\sigma T^3/3, \tag{4.14}$$

where σ is the Stefan-Boltzmann constant.

Despite the fact that the entropy density s_{cond} of condensate is unknown, we can show it in Figure 5 as a set of positive numbers $\lambda=Ts$, if each λ_i is ascribed an equilateral hyperbola

$$s_{cond}=\lambda_i/T.$$

Figure 5 illustrates both curves.

We include the cross-section point c_i of the hyperbola *cd* and the cubic parabola *ab* in Figure 5 in the interval $[c_0, c_k]$. Assume the entropy generation along the line cd outside this interval and the limits of the interval are fixing the boundary of the medium stability.

Absence of the entropy generation inside the interval $[c_0, c_k]$ means that the product $2s_i$ (T_k-T_0) is

$$\int_{T_0}^{T_k} dT(s_{cond}+s_{rad}).$$

By substituting *s* we can see that these equalities are valid only at $T_0=T_k=T_i$. So, if the condensate and radiation are in equilibrium, the equality

$$s_{cond}= s_{rad}= 4\sigma T^3/3 \tag{4.15}$$

is also valid. Thus, when the equilibrium state is achieved the medium extension is realized along the cross-section line of the parabola and hyperbolas. Equalities (4.15) are of fundamental character; all other thermodynamic values for the condensate can be derived from these expressions.

4.6.7. Thermal Radiation with Positive Pressure

In the classical electromagnetisms the internal energy density of thermal radiation is

$$u_{\text{rad}} = 3p_{\text{rad}}, \tag{4.16}$$

In another site, the internal energy density $u(T)$ of thermal radiation is equal to

$$u(T)_{\text{rad}} = \int_0^{\infty} u(\nu, T)\, d\nu = \sigma T^4, \tag{4.17}$$

where $u(\nu, T)$ is the function of the internal energy density depending on frequency ν and temperature T of the radiation. The agreement of theoretical distribution and experimentally observed one $u(\nu, T)$ is found by Planck in the framework of the quantum hypothesis [23,25]. The pressure and the entropy density s_{rad} are

$$s_{\text{rad}}=4\sigma T^3/3, \;\; p_{\text{rad}}= \sigma T^4/3. \tag{4.18}$$

The second law of thermodynamics includes the value u too. It can be written as

$$u = Ts - p, \tag{4.19}$$

where u and s are densities of energy and entropy, respectively. Substituting the values u_{rad}, s_{rad} and p_{rad} from (4.16)-(4.18) in the equation (4.19), one can conclude that the experiment with an ideal radiation does not contradict the classical electromagnetisms, quantum theory and thermodynamics of thermal radiation.

4.6.8. Radiation with Zero and Negative Pressure

The radiator with zero or negative pressure of radiation is unknown. The temperature of radiation in these states is positive and we assume that the internal energy of radiation is equal to σT^4. According to (34.9), (4.19), the internal energy of radiation with entropy S_0 at zero pressure and positive temperature T_α is excessive:

$$u_{p=0} = Ts_0 - 0 = \sigma T^4 + \sigma T^4/3.$$

The radiation should be received energy $\sigma T^4/3$ from the environment at reaching a zero isobar along a surface $\varphi\,(U, S, V) = 0$.

If the initial medium with entropy S_0 reaches zero pressure not along a surface $\varphi\,(U, S, V) = 0$ during the adiabatic process, the energy difference is equal to

$$u_{p=0} - (Ts_0 - p_{\text{rad}}) = 4\sigma T^4/3 - 4\sigma T^4/3 + 0 = 0.$$

In this case the radiation should not be energy received from the environment.

Assume that the primary radiation moves along the curve ***a***(-1) at negative pressure. In this case

$$p^*_{\text{rad}} = -\sigma T^4/3$$

and the internal energy will be

$$u^*_{\text{rad}} = Ts^* - p^*_{\text{rad}} = 5\sigma T^4/3. \quad (4.20)$$

It should be emphasized that the value u^*_{rad} is the internal energy radiation with the negative pressure and radiation is out of the equilibrium with condensate. In this case $u^*_{\text{rad}} > u(T)_{\text{rad}}$. The excessive energy of radiation is equal to

$$\Delta u^* = u^*_{\text{rad}} - u_{\text{rad}} = 2\sigma T^4/3$$

and should be received from the environment for crossing a zero isobar and cooling more low temperature T_a.

Thus, radiation with zero and negative pressure has excessive energy. These states are not reached during the inertial thermal radiation move.

4.6.9. Equilibrium between Thermal Radiation and Condensate

Radiation with negative pressure is instability itself and performs the phase transformation forming the homogeneous equilibrium between radiation with positive pressure and condensate with negative pressure under scenarios *i-j*. According to the equilibrium conditions (4.5), (4.13) and (4.18), the condensate pressure is

$$p_{\text{cond}} = -\sigma T^4/3.$$

This phase transformation carries out at the condition $S_0 = S_{\text{rad}} + S_{\text{cond}}$. As $p_{\text{rad}} = -p_{\text{cond}}$, then

$$u_{\text{rad}} + u_{\text{cond}} = Ts_0.$$

According to (4.16) – (4.18),

$$u_{\text{cond}} = 4\sigma T^4/3 - \sigma T^4 = \sigma T^4/3 = -p_{\text{cond}}.$$

Thus, for the homogeneous phase equilibrium between thermal radiation and condensate the equalities

$$\omega_1 + \omega_2 \neq 0 \text{ at the } u_1 = 3u_2,$$

$$\omega_3 + \omega_4 \neq 0 \text{ at the } u_3 = 3u_4,$$

$$\omega_5 + \omega_6 \neq 0 \text{ at the } u_5 = 3u_6,$$

$$\omega_7 + \omega_8 \neq 0 \text{ at the } u_7 = 3u_8$$

are valid but the curve $\omega = 0$ remains the zero isobar.

Assume that the non-equilibrium initial medium with entropy S_0 reaches the surface φ $(U, S, V) = 0$ in the point with negative pressure. It decays spontaneously according to the scenario 3*j*. According to (4.14), (4.15) and (4.20), the energy

$$u^* - (u_{rad} + u_{cond}) = 5\sigma T^4/3 - 4\sigma T^4/3 = \sigma T^4/3$$

is uncompensated energy which evolves during the phase transformation under scenario 3*j* for the primary radiation instability at negative pressure with the medium consisting of thermal radiation and condensate.

The uncompensated energy can be dissipated in space.

4.6.10. Thermal Radiation and Condensate out of the Equilibrium

The extension of radiation-condensate medium is an inertial process, so that the positive pressure of radiation p_{rad} lowers, and the negative pressure of the condensate p_{cond} increases according to the condition (4.13).

Matter is extended with thermal radiation. As it is known in cosmological theory [23, 25], the plasma inertial extension had led to formation of atoms and distortion of the radiation-matter equilibrium. Further local matter non-homogeneities appeared as origins of additional radiation and, consequently, matter created a radiation excess in the medium after the equilibrium radiation-matter was disturbed. This work supposes that radiation excess may cause equilibrium displacement for the medium, thus radiation and condensate will continue inertial extention in different equilibrium processes in parallel.

We assume that the distortion of the equilibrium radiation-condensate had occurred at temperature T_i of the medium at point c_i in Figure 5. The radiation will be extended adiabatically along the line $c_i a$ of the cubic parabola without entropy generation. While the condition $V_{rad} = V_{cond}$ is satisfied if the equilibrium is disturbed, the equality $s_{cond} = s_{rad}$ points out directions of the condensate extension without entropy generation. As shown in Figure 5, the unchangeable adiabatic isolation is possible if the condensate extends along the isotherm T_i without heat exchange with radiation. Differentiation of the expression

$$U_{cond} - T_i S_{cond} + p_{cond} V = 0$$

With T=const and S=const gives that p_{cond} is also constant.

The medium as a whole extends in such a manner that the positive pressure p_{rad} of radiation decreases, and the negative pressure p^*_{cond} remains constant. As radiation cools down, the dominant p^*_{cond} of the negative pressure arises, and the medium begins to extend with positive acceleration.

5. Experimental Results on ZnCdHgTe and Ag(Tl)-Containing Thin Films Manufactured by Laser Technologies

(100)-oriented monocrystalline *ZnTe* and *CdTe* wafers were selected as substrates. *ZnCdHgTe* films were grown by pulse laser deposition (PLD) technique under dynamic vacuum mode with the following parameters of the *Nd:YAG* laser: wavelength is 1.06 μm, pulse duration is 150 μs, pulse repetition frequency is 10 Hz, and pulse peak power is 0.6 J [36].

The growth chamber has provided a vacuum level of no less than 10^{-5} Pa. The pellets obtained from powdered mixtures of binary compounds *CdTe*, *ZnTe*, and *HgTe* under pressure of 150 technical atmospheres were used as the source for layer deposition [36].

Samples of the as-grown structures with layers of 5.22 μm thickness and (0.3 – 0.5) cm^2 electrical areas were chosen for measurements. All electric studies had been performed under room temperature [36].

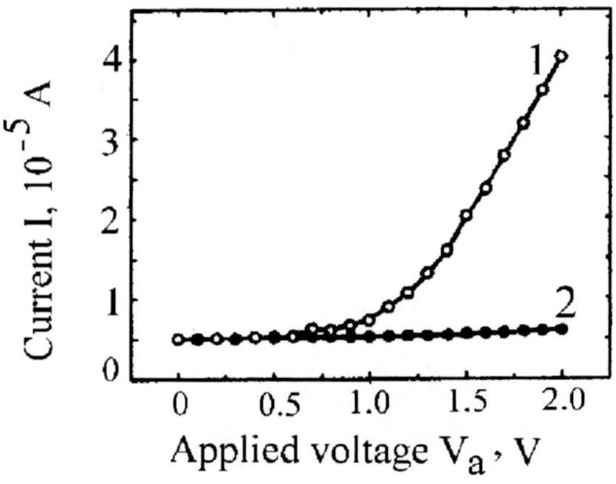

Figure 6. Experimental IVC's of heterostructure *ZnTe/ZnCdHgTe*. T = 290 K. Curve 1 plots the forward current, curve 2 presents the reverse one [36].

Structures with following composition were selected for room-temperature current-voltage measurements: $ZnTe/Zn_{0.08}Cd_{0.37}Hg_{0.53}Te$ and $CdTe/Zn_{0.08}Cd_{0.37}Hg_{0.53}Te$. The intrinsic carrier concentration and energy gap of the as-grown film were estimated as follows [36]:

$$n_i(x,y,T) = \left[6.48 - 4.42x - 6.54y + 1.42\times10^{-3}(1+x+y)\right]10^{14}E_g^{3/4}T^{3/2}\exp\left(-\frac{E_g}{2k_BT}\right) \quad (5.1)$$

$$E_g(x,y,T) = -0.301 + 1.93x + 2.29\times10^{-2}y^{1/2} + 2.731y - 1.62x^2 + 5.35\times10^{-4}T(1 - 2x - 0.35y^{1/2} - 1.28y) + 3.328x^3 - 1.248y^2 + 2.132y^3, \quad (5.2)$$

Calculations were resulted in $E_g(300)$ = 0.86 eV and $n_i(300) = 1.1\cdot10^{11}$ cm^{-3} [36].

Figure (6) shows typical current-voltage characteristics (IVC) of the heterostructure *ZnTe/ZnCdHgTe*. As one can see, the diffusion potential is estimated to be about 1.0 eV [36].

In order to obtain more detailed information about charge carriers' transport through the heterostructure the experimental data were double-log rescaled (Figure (7)). As one can see, IVC of heterostructure *ZnTe(100)/ZnCdHgTe* demonstrates three domains governed by the power law [36].

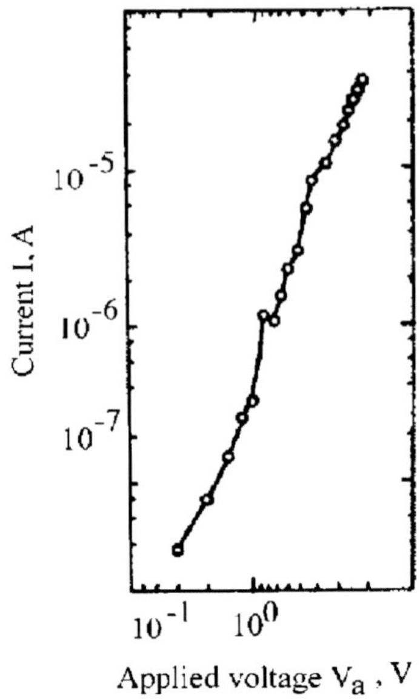

Figure 7. Forward current of the heterostructure *ZnTe(100)/ZnCdHgTe*. T = 290 K [36].

Qualitatively the IVC of this heterostructure is described as follows (here V_a stands for the applied bias) [36]:

$$\begin{cases} I_1 \sim V_a^{1.0},\ 0.0 < V_a < 0.6V, \\ I_2 \sim V_a^{1.4},\ 0.65 < V_a < 1.1V, \\ I_3 \sim V_a^{4.7},\ V_a > 1.15V. \end{cases} \quad (5.3)$$

Analogically studied IVC's of the heterostructure *CdTe/ZnCdHgTe* (Figure (3)) exhibited two similar power-law sections [36]:

$$\begin{cases} I_1 \sim V_a^{1.6}, 0.0 < V_a < 1.06V, \\ I_2 \sim V_a^{2.4}, V_a > 1.1V. \end{cases} \quad (5.4)$$

Qualitative analysis performed according to the classical theories of semiconductor structures has manifested a complete current mechanism in both types of heterostructures.

In particular, the ballistic mode of the carriers transfer in the heterostructure *CdTe/ZnCdHgTe* is changed by the tunneling-recombination regime; for the heterostructure *ZnTe/ZnCdHgTe* we have observed three modes (see (5.3)): velocity saturation mode turns into a ballistic one and this regime is replaced by the strong tunneling (trap-assistant in the space-charge region) of the carriers.

Capacitance-voltage measurements carried out at 290 K under test signal frequency f = 1 kHz were shown no barrier features for *ZnTe/ZnCdHgTe*heterostructure. On the contrary, the heterostructure*CdTe/ZnCdHgTe* has revealed CVC typical for graded-gap structures with step-like distribution of the charged centers in the space-charge region (Figure (9)). As one can see, an IR-photosensitivity is observed: diffusion potential changes from 0.20 eV (dark CVC) to 0.50 eV (light) under 0.75 μm light source illumination [36].

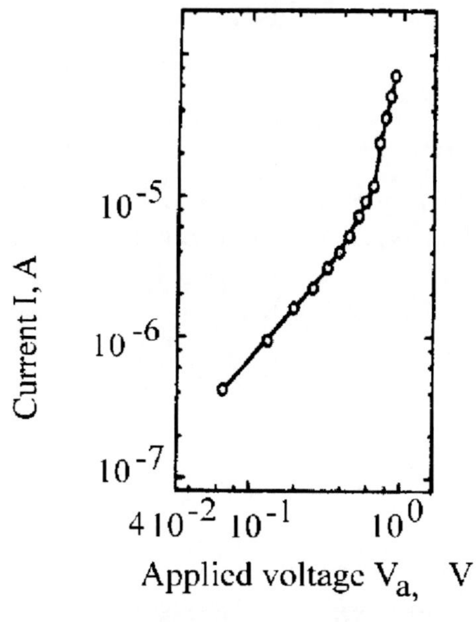

Figure 8. Forward IVC of the heterostructure *CdTe(001)/ZnCdHgTe*. T = 290 K [36].

The experimental data obtained from the C-V-measurements have made it possible to estimate some parameters of the heterostructure*CdTe/ZnCdHgTe* [36]: the space charge region capacitance C_{ss}, the widths of the space charge region W_1 (from *CdTe* side) and W_2 (from the *ZnCdHgTe* side), the density and energy spectrum of the charge centers localized at the interface (N_{ss} and E_{ss}, respectively) [36]. Results obtained from the calculations are listed in Table 1 [36].

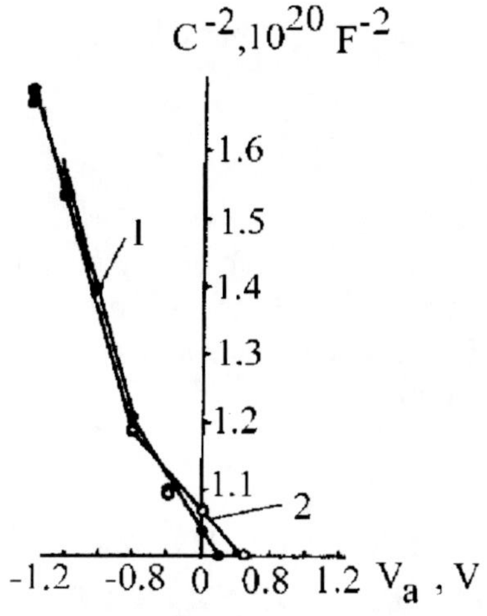

Figure 9. CVC (capacitance-voltage characteristic) of the heterostructure*CdTe/ZnCdHgTe*: T = 290 K, *f* = 1 kHz; curve 1 plots CVC under darkness, curve 2 plots CVC under the sample illumination by 0.75 μm source [36].

Table 1. Room-temperature electrical parameters of the heterostructure $CdTe/Zn_{0.08}Cd_{0.37}Hg_{0.53}Te$ [36]

C_{ss}, $F{\cdot}m^{-2}$	W_1, μm	W_2, μm	N_{ss}, cm^{-2}	E_{ss}, eV
$1.19{\cdot}10^{-6}$ dark $7.53{\cdot}10^{-7}$ illumination	58.6, dark 92.6, illumination	$6.1{\cdot}10^{-3}$, dark $9.7{\cdot}10^{-3}$, illumination	$6.1{\cdot}10^{8}$, dark $9.7{\cdot}10^{8}$, illumination	0.54, dark 0.51, illumination

The experimental results may be interpreted as follows: it is known that an equivalent electric circuit of such a diode-like structure involves three resistors: the first two ones, R_j related to the junction (space charge region) and R_{sh} related to the leakage currents in shunt with R_j, are in parallel. The third, resistor R_s, R_s is a series resistor. So, the current-voltage relation can be written as [36]

$$I = (V_a - R_s I)(R_j + 1/R_{sh}). \tag{5.5}$$

Let's go back to the data plotted in Figure (7) and Figure (8). The first domain of IVC for heterostructure*ZnTe/ZnCdHgTe* (5.3) corresponds to leakage currents (up to $8{\cdot}10^{-8}$ A) and determines a possible range of the voltage increase. Unlike this structure, the heterojunction*CdTe/ZnCdHgTe* reveals no considerable leakage currents. The next two domains for both structures can be expressed as follows [36]:

$$J = A n_i^2 \exp\left(\frac{eV_a}{m_1 k_B T}\right) + B n_i \exp\left(\frac{eV_a}{m_2 k_B T}\right), \tag{5.6}$$

where *A* and *B* are terms determining tunneling-recombination processes in the space charge region localized at the interface of the heterojunctions, m_1 and m_2 take care of exponents in (5.3), (5.4), k_B is the Boltzmann constant, *T* is the temperature of experiment (in our case room temperature). Both terms *A* and *B* are functions depending on the tunneling probability T_{tun}, the barrier height Φ_b and parameters of narrow-gap films, first of all on effective mass and mobility of carriers [36]:

$$\{A, B\} \propto f(m^*, \mu)\exp(e\Phi_b/k_B T). \tag{5.7}$$

These results give a possibility of technological governing electric and photo-electric behavior of the investigated structures by means of different doping (we should note that all structures described here are intrinsic semiconductors) of the epitaxial layers and changing the composition (*x*, *y*) of the solid solution [36].

6. Formation Of Barrier Structures in Epitaxial *ZnCdHgTe* Layers by Laser Irradiation

The new narrow gap semiconductor *ZnCdHgTe (ZMCT)* seems to be a good alternative to the worldwide known and used *CdHgTe (MCT)*. Both compounds are solid solutions but the quaternary solid solution *ZnCdHgTe* reveals more stable properties due to the stabilizing effect of *Zn-Hg* bond in comparison with MCT [31]. Besides good promising optoelectronic properties, the new semiconductor is an attractive compound for the examination of various electric field-induced phenomena due to formation of *ZnCdHgTe/substrate* heterostructures [36]. Furthermore, such a heterostructure seems to be a very interesting object for numerical modeling of electrical characteristics determined by the conditions of the growth technology. This part of the Chapter 6 is focused on the room-temperature electric studies performed for the isotype*p-p*-structures $Zn_xCd_yHg_{1-x-y}Te/CdTe$after CO_2-laser radiation. Results described here are obtained for the first time [36].

Table 2. Parameters of the *ZnCdHgTe/CdTe*heterostructure [36]

p-$Zn_{0.18}Cd_{0.12}Hg_{0.7}Te$	CdTe
BandgapE_g, 0.35 eV	BandgapE_g, 1.44 eV
Electron effective mass m_e, 0.02 m_0	Electron effective mass m_e, 0.12 m_0
Charge carriers concentration $5\cdot 10^{16}$ cm^{-3}	Charge carriers concentration ~10^{14} cm^{-3}
Lattice constant a, $6.44\cdot 10^{-12}$ m	Lattice constant a, $6.48\cdot 10^{-12}$ m
Hole effective mass m_h, 0.55 m_0	Hole effective mass m_h, 0.40 m_0

$Zn_xCd_yHg_{1-x-y}Te$ *(0<x≤0.18, 0<y≤0.07)* epilayers of *p*-type conductivity with effective hole concentration p~10^{16} $cm^{-3}$and mobility μ~10^4 cm^2/Vs grown by the modified liquid phase epitaxy (LPE) method on *(111)-CdTe<Zn>*substrates had been chosen as an object of the study [31].

Previous electrophysical and photoelectric measurements revealed the effective hole concentration $5\cdot 10^{16}$ cm^{-3} and carriers life-time ~10^{-6} s. The examined samples (their main parameters are listed in Table 2) were irradiated by CO_2-laser from the ZCMT side in air at room temperature (λ=10.6 μm). Electrical investigations (I-V, C-V) carried out immediately after samples grown at T = 300 K showed no peculiarities. Studies performed after laser irradiation of the samples (Figure (9)) demonstrated that the C-V and I-V dependencies are similar to the ones of barrier structure, and appeared under very weak external electric fields (no more than 200 V/cm) [36].

The experimental current-voltage (I-V) characteristics are plotted in Figure (10). As shown, both sections of the forward I-V dependence are of exponential character:

$$I_{exper} \sim \exp\left(\frac{eV_a}{nk_BT}\right), \tag{6.1}$$

where *e* is the elemental charge, k_B is the Boltzmann constant, V_a is an applied voltage, *n* is the non-ideality factor determined by the charge carriers transport mechanism [36], n~1 for section I and $n<1$ for section II. The detailed numerical analysis demonstrated that the data obtained under current-voltage measurements could be described as follows [36 and Refs. therein]:

$$I = D_{tun} A_{el} \frac{e n_i W}{\tau} \frac{k_B T}{\varphi_0 - eV_a} \exp\left(\frac{eV_a}{k_B T}\right), \tag{6.2}$$

where D_{tun} is the barrier transparency coefficient (see below), A_{el} is an electrical area of the sample, n_i is the intrinsic carrier concentration determined in assumption of *ZnCdHgTe* band parabolicity, W is the space charge width (determined from the capacitance-voltage studies as well as φ_0 appeared as the height of the potential barrier formed due to laser irradiation), τ is the effective lifetime of the charge carriers [36].

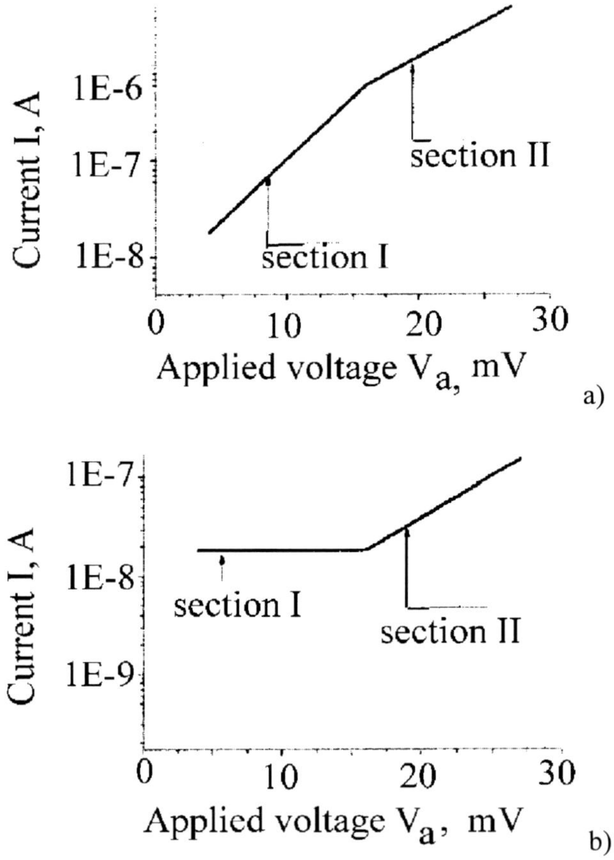

Figure 10. Forward (a) and reverse (b) current-voltage characteristics of the investigated structure after CO_2-laser radiation [36].

Reverse current (Figure (10a)) is not sufficiently governed by the applied bias up to 0.017 V; as the applied voltage increases, the observed current is given by equation (6.1):

$$I = A_{el} \frac{e n_i W}{\tau} \frac{\left((E_g/2) - V_a\right)^{1/2} - \left(E_g/2\right)^{1/2}}{\left(\varphi_0 - eV_a\right)^{1/2}}, \tag{6.3}$$

where E_g stands for the epilayer gap [36].

Capacitance-voltage (C-V) studies performed after laser irradiation of the sample have shown typical Schottky-barrier character (Figure (11)), in other words, the dependence $C^{-2} = f(V_a)$ is sublinear. The barrier height is estimated to be $V_d \approx 0.12$ eV; experimental C-V data make it possible to calculate the main parameters of the formed barrier structure (Table 3) [36].

Table 3. Parameters of the space charge region in quasi-heterostructure *p-p-ZnCdHgTe* [36]

Width of the space charge region W, μm	0.066
Screening length L_{Ds}, μm	0.015
Surface electric field F_s, V/m	$3.65 \cdot 10^6$
Charge centers concentration N_{cc}, $eV^{-1}cm^{-2}$	$2.74 \cdot 10^{12}$
Capacitance of the space charge region C_{cc}, F	0.002

Here are the main expressions used for calculations:

$$W = \left(\frac{2\varepsilon \varepsilon_0 V_d}{eN_a} \right)^{1/2}, \quad L_{Ds} = \left(\frac{\varepsilon \varepsilon_0 k_B T}{2e^2 N_a} \right)^{1/2},$$

$$F_s = \frac{k_B T}{e} \sqrt{\frac{V_d}{L_{Ds}}}, \quad N_{cc} = \frac{1}{e} \sqrt{\frac{2\varepsilon \varepsilon_0 e N_a}{V_d}} \quad \text{and}$$

$$C_{cc} = \frac{\varepsilon \varepsilon_0}{W}.$$

The surface morphology [36] of the samples after laser irradiation has changed: cell-like clusters form the quasi-two-dimensional percolation grid [36]. We suppose that under weakly applied electric fields this grid behaves as an aggregate of *source-drain* type micro-structures, where cell boundaries seem to be a source of the carriers. Thus, the current-voltage characteristic as a visualization of carriers transport processes is effected by the transparency of the potential barrier formed (in our case $D_{tun} \sim 0.02$). These phenomena are of particular interest, especially under a very small applied bias. Figure (12) demonstrates the theoretical I-V characteristics [36].

Comparison of calculated and experimental data (Figure (10) and Figure (11)) shows a good agreement between qualitative and quantitative character of the forward section I (Figure (10a) and Figure (11a)); section II of the experimental reverse current (Figure (10a)) also has similar characteristics with the same theoretical dependence (Figure (11b)); however, the initial section of the calculated reverse current is different from the experimental. This fact indicates the presence of a strong shunting surface resistance, which appeared after irradiation [36 and Refs. therein].

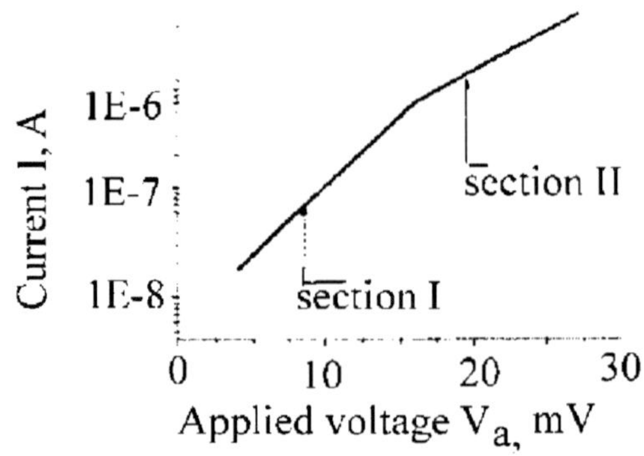

a)

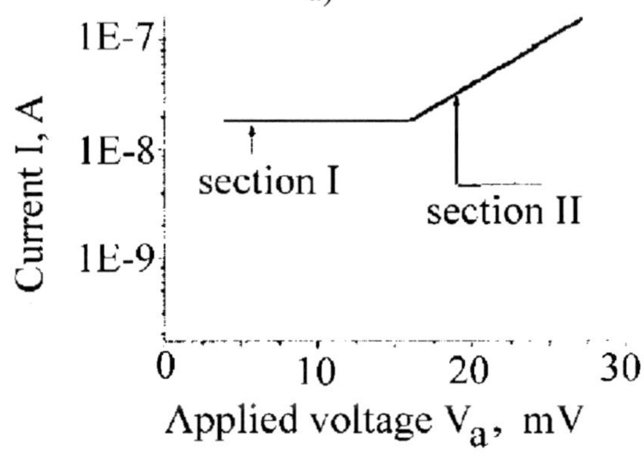

b)

Figure 11. Theoretical I-V-characteristics of the CO_2-laser irradiated *ZnCdHgTe* epilayer [36].

7. Study of Functional Possibilities of Active Elements Based on *ZnCdHgTe* Films Obtained by Laser Technology

Narrow-gap semiconductor materials belonging to the group A^2B^6 compounds are important for some fields of optoelectronics used for various needs of device fabrication. Heterojunctions, based on epitaxial layers *ZnCdHgTe,* are of particular interest owing to the simplicity and reliability of growth technology (modified liquid phase epitaxy and pulse laser deposition (PLD) technique).

The investigation of experimental data, obtained from the simplest electric-field measurements, provides a reasonable base for the computational modeling of active elements based on graded-gap semiconductors [36].

Monocrystalline $Zn_xCd_yHg_{1-x-y}Te$ ($0<x<0.49$, $0<y<0.18$) films were grown on glass substrates by PLD technology using pulse YAG:Nd laser. Deposition was carried out onto substrates under room temperature. Thickness of the layers was estimated to be 0.065–0.70 μm, room-temperature resistance of the as-grown layers was estimated to be about 10^7 Ohm [36].

As the targets for the deposition, pellets prepared from powder-like binary compounds *CdTe*, *ZnTe* and HgTe were used. To obtain a regular composition of the grown layers the targets were rotated around their axis of symmetry with the distance between the target and substrate of 60–90 mm; deposition time ranged from 2 to 6.5 h, the growth chamber pressure was about 10^{-4} Pa. Room-temperature parameters of the as-grown films are as follows: $E_g(Zn_{0.49}Cd_{0.07}Hg_{0.44}Te)$ = 0.81 eV, $E_g(Zn_{0.10}Cd_{0.10}Hg_{0.80}Te)$ = 0.24 eV, $E_g(Zn_{0.39}Cd_{0.18}Hg_{0.43}Te)$ = 0.86 eV, and $n_i = 3.1 \cdot 10^{11}$ cm^{-3}, $n_i = 1.1 \cdot 10^{16}$ cm^{-3} and $n_i = 1.4 \cdot 10^{11}$ cm^{-3}, respectively.

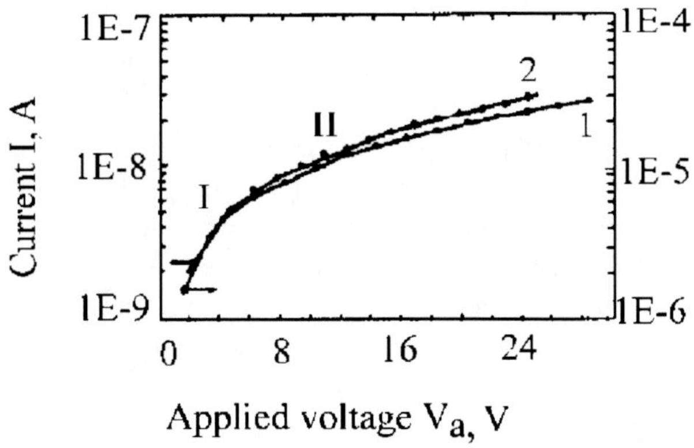

a)

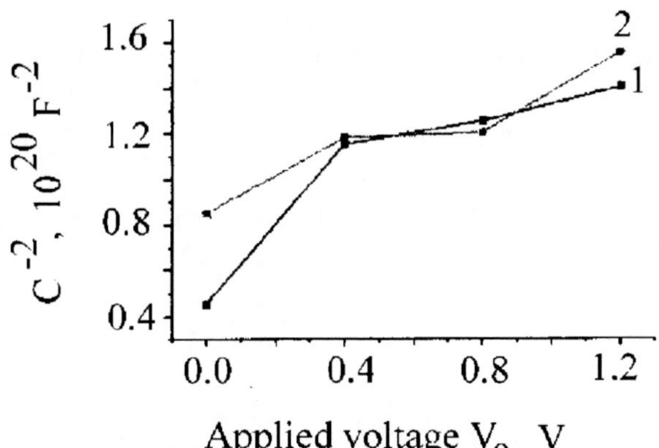

b)

Figure 12. (a) Current–voltage characteristics of the investigated samples, T = 290 K. • - dark current, ○ -light current. Slope I is 2.4, slope II is in the range 0.5-1.07. (b) Capacitance–voltage characteristics of the heterostructure *ZnCdHgTe/CdTe* [36].

The best samples of the 1.8 mm^2 electric area A_{el} were chosen for studies of the room-temperature current-voltage characteristics. The measurements were carried out also under the illumination of the samples by the light source of 100 W power. The range of applied bias V_a was 0-30 V. Note that the experimental current-voltage characteristics demonstrated no considerable difference regardless of the various values of the gap [36].

Figure (11a) plots the typical experimental data. As one can see, these dependencies are of diode-like character, so that we are able to simulate the data according to the classical models of the physics of semiconductor devices [36 and Refs. therein].

To investigate the photosensitivity of the films grown on the single crystal *CdTe* substrates capacitance–voltage measurements were carried out for the as-grown heterostructure *ZnCdHgTe/CdTe* (Figure (11b)). Both dark and light dependencies are also of the barrier-like character [36].

As shown, both dark and illuminated current–voltage dependencies are similar. The weak additional photocurrent is observed under the irradiation of the films by the source of incoherent light. The threshold bias signing the change of the carriers transport mechanism is about 1 V. The current–voltage characteristic can be described as follows [36 and Refs. therein]:

$$I_r = I_s\exp(eV_a/nk_BT) + XA_{el}N_0(D_n/\tau_n)^{1/2}\exp[-(eV_d + E_i)/k_BT](\exp[eV_a/k_BT] - 1), \quad (7.1)$$

$$X = \exp[-(4(2m^*)^{1/2}/3\hbar)d(U-V_d)^{3/2}/U + eV_d/k_BT)], \quad (7.2)$$

where n is the slope of the experimental curve, X is the tunneling transparency coefficient strongly dependent on the threshold voltage U (in our case X is calculated according to the parabolic VCB approximation), N_0 and V_d are the concentration of the charge carriers at the top of the potential barrier and its height (these values are usually determined by the capacitance-voltage studies), d is the barrier width, D_n and τ_n take care of the electrons diffusion coefficient and lifetime, respectively, E_i is the semiconductor potential energy required by the charge carrier to escape the barrier top, which also appeared as a function of the .lm surface relief and the applied voltage [36 and Refs. therein].

The numerical analysis of the experimental data revealed the necessity of taking into account the effect of atmospheric oxygen (at least up to one monolayer) on the experimental current–voltage characteristics. Due to the large electronegativity (≈ 3.5 eV), oxygen has to create acceptor states in the corresponding material. Their filling (i.e. the dipoles formation) occurs during the process of electrons transfer from the valence band of the compound. The numerical simulation has demonstrated that the current-voltage characteristics performed for the prepared film (Figure (13)) are to be considered for two ranges of the applied voltage: (i) 0-0.4 V and (ii) up to 20 V. The initial section (under V_a up to 0.4 V) is described by the function given below [36]:

$$I_{theor}=I_sXA_{el}eN_1\exp(-\Delta E_v/k_BT)(D_n/\tau_n)^{1/2}\exp[(eV_a/k_BT) - 1], \quad (7.3)$$

where N_1 is defined from the low frequency (f = 1 kHz) capacitance-voltage measurements. As the applied bias increases, the current is

$$I_{theor(2)}=I_{theor}\exp(-e\Delta_{tun}/k_BT), \quad (7.4)$$

here$D_{tun} = I_{ZnTe} - I_{CdTe}$, I is the electronegativity of the corresponding compound.

For irradiated *ZnCdHgTe* films the current, observed experimentally, may be written as follows [36 and Refs. therein]:

$$j = j_0[\exp(eV_a/nk_BT) - 1] - j_{sc}, \quad (7.5)$$

j_0 is a thermionic emission current, and j_{sc} is the short circuit current. The open circuit voltage is [36]

$$V_{oc} = (nk_BT/e)\ln[(j_{sc}/j_0)+1], \tag{7.6}$$

Where $j_0 = A^*T^2exp(-e\varphi_b/k_BT)$ and $A^* = 4\pi m^*e(k_B)^2/\hbar^3$.

Results of the numerical simulation are plotted in Figure (13) [36].

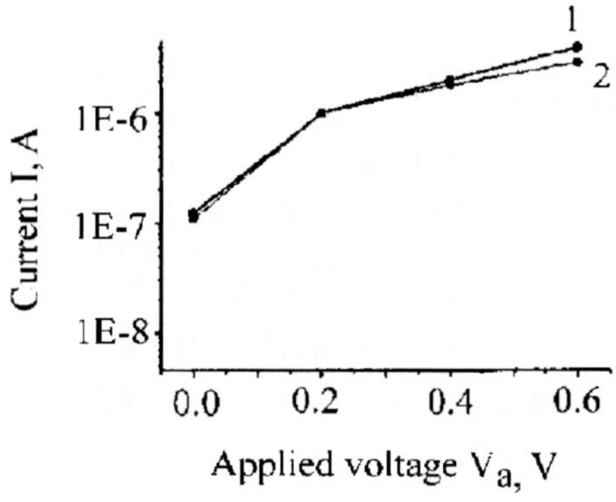

Figure 13. Results of the numerical experiment for the *ZnCdHgTe* film (1 - experiment, 2 - theory) [36].

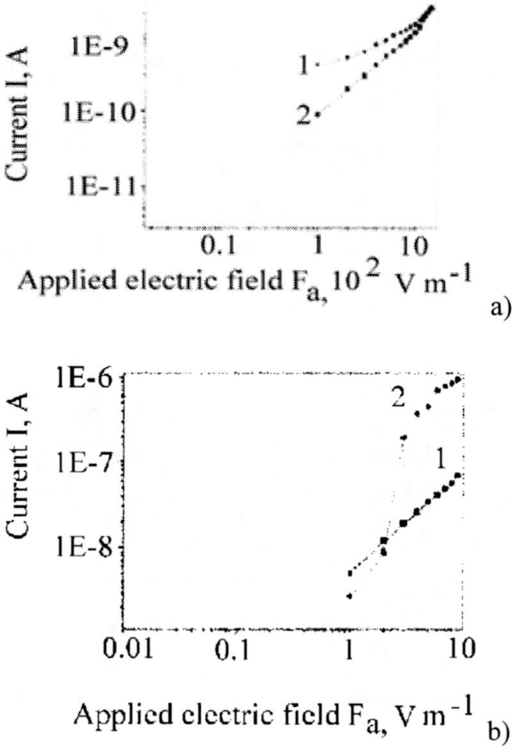

Figure 14. Current-field characteristics of the structures In/Tl_3SbS_3 (a) and In/Ag_3SbS_3(b) [36].

Chalcogenide semiconductors are also of special importance for a novel nanolectronics. Thin films of the corresponding composition with thickness up to 140 nm were obtained by means of pulsed laser deposition technology onto *KC*l substrates at room temperature.

Scanning electron microscopy (SEM) studies have revealed the inhomogeneous structure of the deposited films [36].

Experimental field characteristics of the investigated samples are plotted in Figure (14). Numerical analysis of the data obtained has shown the complete law function governing charge carriers current flowing through the structure. It was found that the MS-structure was characterized by current-field dependency $I \sim (F_a)^m$, where m varies from 1 to 4 whereas the applied electric field increases. Structures In/Tl_3SbS_3 are characterized by $m \sim 0.4\text{-}1.0$ in the whole range of the applied field [36].

As one can see, both structures demonstrate a space-charge limited (SCL) current, but the MS In/Tl_3SbS_3 have a weak SCL dependence as well as a reverse current of MS In/Ag_3SbS_3; the forward current-field function of the structure In/Ag_3SbS_3 shows a sound SCL current under the conditions of the experiment [36].

The current-voltage (IVC) characteristics of the Ag_3SbS_3 films are shown in Figure (15) (for more detailed analysis the experimental data were rebuilt in double-logarithmic scale) [36].

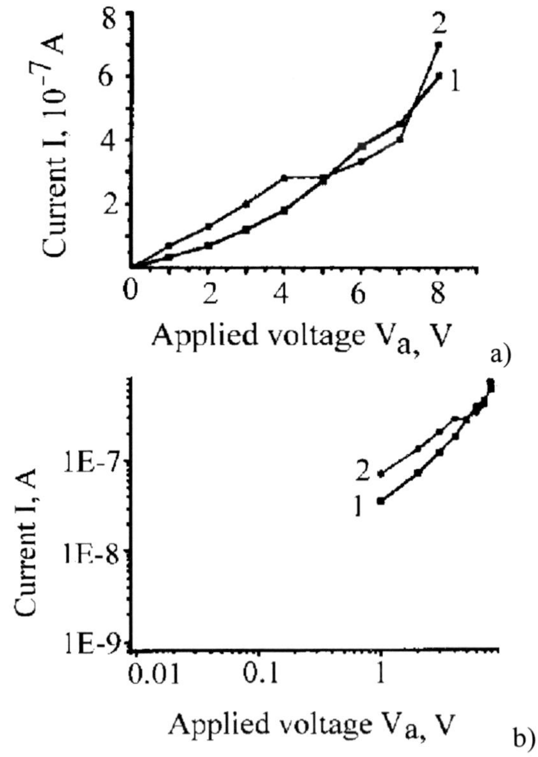

Figure 15. IVC of the pulse laser deposited film Ag_3SbS_3: experiment (a) and double-log recalculation (b) [36].

As one can see, both forward and reverse sections of the experimental IVC are of weak exponential character and qualitatively can be described as $j \sim (V_a)^m exp(eV_a/mk_BT)$, where V_a is an applied voltage, $m \sim 1$. Both forward and reverse sections of the experimental IVC show the broken path due to the asymmetrical transport of the charge carriers under the applied bias. The strongly inhomogeneous structure of the film (Figure (16) leads to the considerable carriers transfer through the net of the barriers with disordered values of the height [36]. The next part of this chapter deals with the calculation of the energy spectrum of the interface charge centers (ICC) and their effect on the current; as the test the MS In/Ag_3SbS_3 was chosen [36].

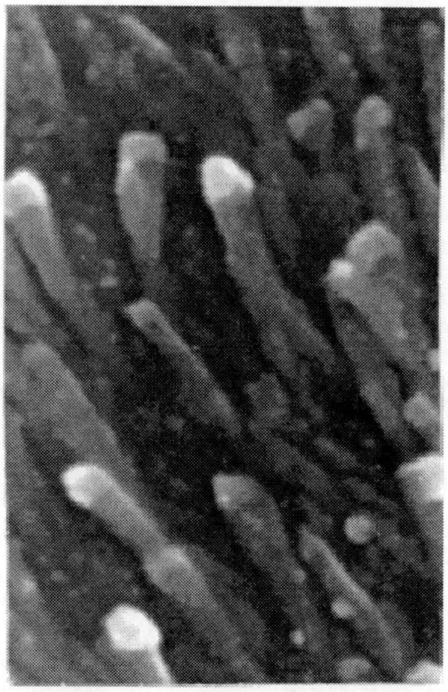

Figure 16. Typical SEM image of the investigated Ag_3SbS_3 film [36].

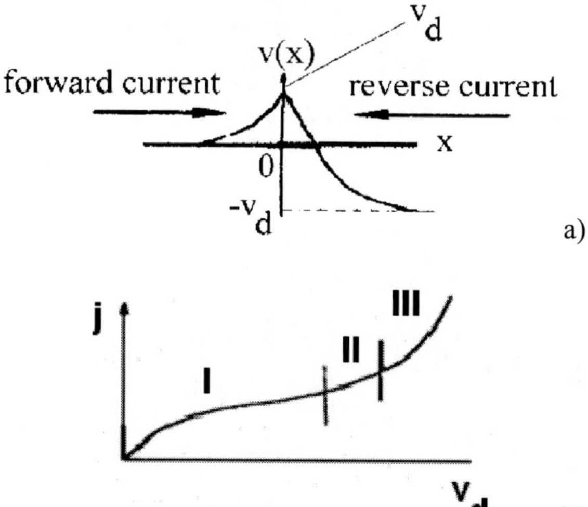

Figure 17. Band scheme of the semiconductor in the interface vicinity at $V_a \neq 0$ (a) and qualitative I-V characteristic (b); modes of the barrier filling: I – weak, II – intermediate, III – limit [36 and Refs. therein].

The following consideration [36] is based on the comparison of the theoretical data calculated in assumption of a weak filling of the top of potential barrier by the ICC and the experimental results [36].

The electron states with the concentration N_s and filling degree n_s, exist at the plane $x = 0$ (Figure (17)) [36 and Refs. therein].

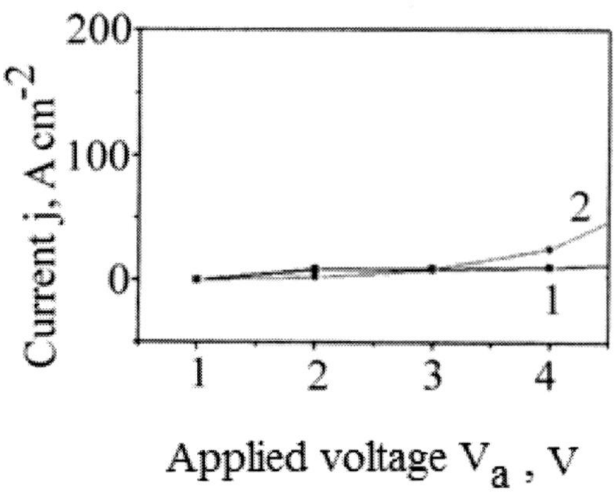

a)

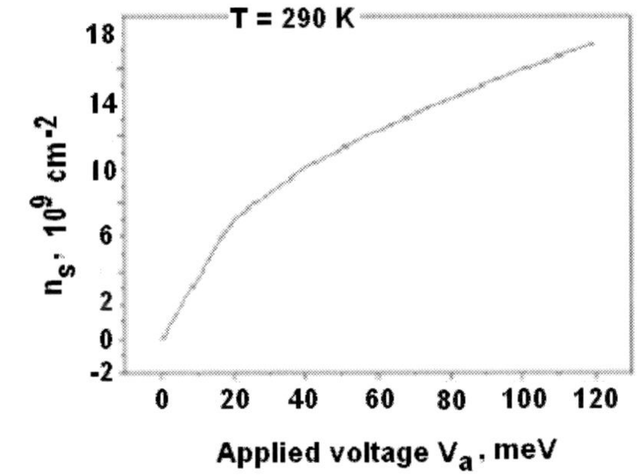

b)

Figure 18. Experimental (curve 1) and theoretical (curve 2) current (a) and surface charge states concentration at T = 290 K (b). A good agreement is observed under the applied voltage up to 50 meV [36].

The current density j and the electron concentration n_0 at the top of the barrier are:

$$j = -\frac{\mu k_B TN}{L_D}\sqrt{\frac{v_s}{2}}\left[1-\left(\frac{v_d^2}{16v_s^2}\right)\right]\exp(-v_0)(1-\exp(-v_d)), \tag{7.7}$$

$$n_0 = \frac{N}{2}\left(1-\frac{v_d}{4v_s}\right)\exp(-v_0)(1-\exp(-v_d)), \tag{7.8}$$

where μ is the charge carriers mobility (this value is determined from the experimental electric measurements), N stands for the carriers concentration in the bulk of the structure (i.e., in the bulk of the Ag_3SbS_3 monocrystal), which are assumed to be completely ionized [36]. Here [36 and Refs. therein]:

$L_D = \sqrt{\dfrac{\varepsilon_0 \varepsilon k_B T}{e^2 N}}$ is the Debye length,

$$v_s = \frac{n_s^2}{8N L_D^2},\ v_d = \frac{eV_a}{k_B T},$$

$v_0 = \dfrac{e\phi_0}{k_B T}$ is a dimensionless barrier for the forward current, φ_0 is determined from the capacitance-voltage measurements, and

$$v_0 = -\frac{1}{v_s}\left(v_s - \frac{v_d}{4}\right)^2. \tag{7.9}$$

Results of calculations are presented in Figure (18) [36].

CONCLUSION

Let us note the following:

There are no experimental facts that allow ascribing the negative pressure to some ions, proteins, their groups or water as the cytosol components. However, it is known [19, 20] that the osmotic pressure with an opposite sign is applied to explain the transportation of substances in a plant cell. We hope that the phenomenological investigation of the cytosol from the living cell as a model from the first thermodynamics principles will find a niche among theory and experiment for explanation of mechanisms of biochemical processes in the living cell and their realization in novel biotechnologies. Analysis of the cytosol phase network requires a combination of experimental and theoretical approaches, including the development and analysis of simulations and modeling.

The negative pressure of the condensate of thermal radiation is Pascal-like and isotropic; it is constant from the moment that the equilibrium with radiation was disturbed by the condensate and is equal (by absolute value) to the energy density. The condensate of thermal radiation is a physical medium, which interacts only with the radiation and this physical medium penetrates the space as a whole. This physical medium cannot be obtained under laboratory conditions because there are always external forces for a thermodynamic system in the laboratory. As known [33], the photon Bose-condensate is obtained. This condensate has no negative pressure while it is localized in space. It seems very interesting to find, in nature, a condensate of thermal radiation with negative pressure. Possible forms of physical medium with negative pressure and their appearance at cosmological observations are widely discussed [34]. The radiation can consist of other particles, and then the photon, among them may also be unknown particles. We hope that modeling the medium from the condensate and radiation will be useful for checking the hypotheses and will allow explanation of the nature of the substance responsible for accelerated extension of the Universe. The medium from thermal radiation and condensate is the first indication of the existence of physical vacuum as one of the subjects in classical thermodynamics and the complicated structure of the dark energy [35].

Electrical characteristics (current-voltage and capacitance-voltage) of the heterostructures based on epitaxial films of narrow-gap semiconductor solid solution *ZnCdHgTe* and monocrystals*CdTe* and *ZnTe* are studied at room temperature. The investigations demonstrated typical barrier properties for all structures. Room-temperature IR-photosensitivity is revealed by the heterostructure $CdTe/Zn_{0.08}Cd_{0.37}Hg_{0.53}Te$ in the wavelength range 0.6-0.7 μm; considerably low density of charge centers localized at the interface of the heterostructure $CdTe/Zn_{0.08}Cd_{0.37}Hg_{0.53}Te$, exhibited by C-V-studies, makes it possible to use such semiconductor structures for prospective IR-photonics applications [36].

The epilayer surface studies had demonstrated the cell-like structure replacing the original tessellated structure observed on the as-grown samples where cell boundaries are supposed to be regions of free electron accumulation. Experimental electrical characteristics are typical for the structures with a space-charge region. These experimental results seem to be a base for practical use of the laser treatment technique of lasers to produce artificially defected surfaces of the thick semiconductor films [36].

Monocrystalline thin films of the narrow-gap semiconductor *ZnCdHgTe* grown by PLD technology are investigated for the first time. The room temperature electric studies (in particular, current–voltage measurements) revealed the diode-like characteristics. The additional photocurrent, caused by noticeable light modulation of the minority charge carriers, is observed under irradiation by the light source of 100 W power [36].

The room-temperature electric field – current characteristics are investigated for the thin films and Schottky-like MS-structures In/Ag_3SbS_3. The sufficiently asymmetrical carriers transport is registered for the investigated films due to the inhomogeneous surface structure. The numerical experiment presented here shows the considerable effect of the applied electric field-dependent filling of the potential barrier localized at the metal-semiconductor interface under V_a= 0-50 meV. As the applied bias is increased, the tunneling current becomes a dominant component, which is observed experimentally. At the same time, the tunneling through the intercrystallite boundaries, appears to be significant at $V_a \leq E_g/2$ [36].

REFERENCES

[1] Sydorchuk, P., Khlyap, G., Andrukhiv, A., *Cryst. Res. Technol.*, 36 (5-6), pp.361-369, 2001.

[2] Khlyap, G., Sydorchuk, P., *Cryst. Res. Technol.*, 36 (8-10), pp.1027-1034, 2001.

[3] Sze, S., *Physics of Semiconductor Devices*, John Wiley and Sons: New York, pp. 4-37, 1981

[4] Gulyamov, G., Umarov, K. B., *Semiconductors (Russia)*, 29 (1), pp. 33-38, 1995.

[5] H. Lodish, A. Berk, P. Matsudaira, C. A. Kaiser, M. Krieger, M. P.Scott, L. Zipursky, J. Darnell, *Molecular Cell Biology*, 5th ed. (W. H. Freeman, 2004).

[6] L. Lehninger, D. L. Nelson, and M. M. Cox, *Principles of biochemistry*, 5rd ed. (Freeman, New York, 2008).

[7] D. J. Voet, J. G. Voet, and C. W. Pratt, *Principles of biochemistry*, 3rd ed. (John Wiley and Sons, New York, 2008).
[8] J. M. Berg, J. L. Tymoczko, and L. Stryer, *Biochemistry*, 6rd ed. (Freeman, New York, 2006).
[9] J. W. Gibbs, *The Scientific Papers of J. W. Gibbs, Thermodynamics*, (Kessinger Publishing, New York, 2007), Vol. 1.
[10] I. Prigogin and D. Kondepudi, *Modern Thermodynamics* (John Wiley and Sons, New York, 1999).
[11] Muenster, *Classical Thermodynamics* (Wiley-Interscience, New York, 1971).
[12] P. Bazarov, *Thermodynamics* (Pergamon Press, Oxford, 1991).
[13] C. Goodwin, *Temporal Organization in Cells* (Academie Press, London-New York, 1963).
[14] N. S. Pechurkin, *Energy and Life* (Nauka, Novosibirsk, 1988).
[15] E. P. Odum, *Basic ecology* (CBS College Publishing, New York, 1983).
[16] P. Ivanov, Usp. Sovrem. Biol., Vol. 128, No. 6, 606 (2008).
[17] C. Kittel and H. Kroemer, *Thermal Physics* (Freeman, New York, 2000).
[18] V. I. Laptev, Dokl. *Phys. Chem.*, Vol. 429, part 2, 243 (2009); Vol. 430, part 2, 36 (2010).
[19] N. P. O. Green, G. W. Stout, D. J. Taylor, and R. Soper, *Biological science* (Cambridge University Press, Cambridge, 1997).
[20] P. H. Raven, R. F. Evert, and S. E. Eichorn, *Biology of plants* (8rd ed., Freeman, New York, 2012).
[21] W. Gibbs, *The Scientific Papers of J. W. Gibbs, Thermodynamics*, (Kessinger Publishing, New York, 2007), Vol. 1.
[22] C. Kittel and H. Kroemer, *Thermal Physics* (Freeman, New York, 2000).
[23] Prigogin and D. Kondepudi, *Modern Thermodynamics* (John Wiley and Sons, New York, 1999).
[24] Muenster, *Classical Thermodynamics* (Wiley-Interscience, New York, 1971).
[25] P. Bazarov, *Thermodynamics* (Pergamon Press, Oxford, 1991).
[26] A.S. Kompaneets, *Soviet Physics - JETP*, Vol. 4, No. 5, 730 (1957).
[27] H. Dreicer, *Phys. Fluids*, Vol. 7, No. 5, 732 (1964).
[28] R. Weymann, *Phys. Fluids*, Vol. 8, No. 11, 2112 (1965).
[29] Ya.B. Zel'dovich and R.A. Syunyaev, *Soviet Physics - JETP*, Vol. 35, No. 1, 81 (1972).
[30] A.E. Dubinov, Technical Physics Letters, Vol. 35, No.3, 260 (2009).
[31] V.A. Kuz'min and M.E. Shaposhnikov, *JETP Lett.*, Vol. 27, No. 11, 628 (1978).
[32] G. Spavieri and M. Rodriguez, Phys. Rev. A, Vol. 75, 052113(2007).
[33] J.Klaers, J. Schmitt, F. Vewingerand M. Weitz, *Nature*, Vol. 468, 545 (2009).
[34] P. J. E. Peebles and Bharat Ratra, *Reviews of Modern Physics*, Vol.75 (2), 559 (2003).
[35] V.I. Laptev and H. Khlyap, in *Solar Cells: new aspects and solution*, edited by L.A. Kosyachenko, (INTECH, Rijeka, 2011), p. 357.
[36] H. Khlyap, Physics and Technology of Semiconductor Thin Film-Based Active Elements and Devices. Bentham Science Publishers Ltd., 2009, 127 pp. eISBN: 978-1-60805-021-5.

In: Laser Technology, Applications and Future Prospects
Editors: D. Ebersbach and F. Gottschalk

ISBN: 978-1-62417-088-1

Chapter 3

PREDICTION AND ANALYTICAL DESCRIPTION OF SINGLE LASER TRACKS GEOMETRY. CHARACTERIZATION AND ANALYSIS OF 316L STAINLESS STEEL MICROSTRUCTURE

Hussam El Cheikh and Bruno Courant
LUNAM Université, Université de Nantes, Ecole Centrale Nantes,
Institut de recherche en Génie Civil et Mécanique (GeM)
(UMR CNRS 6183) Equipe Etat Mécanique et Microstructure (EMM),
CRTT Saint-Nazaire cedex, France

ABSTRACT

Direct Laser Fabrication (DLF) is a process making possible the manufacture of functional parts directly from a command file injecting powder in a laser beam. The powder melting and solidification processes lead to the fabrication of this part layer by layer. In this study, deposition of 316L stainless steel powder on a steel substrate is carried out using a 700W fiber laser. Three values of each processing parameters (laser power P, scanning speed V and powder mass flow Q_m) are fixed and so 27 different experiments have been made and analyzed. The layer geometry is an important process characteristic and its mastery is essential to control the final part fabrication. Analytical relationships between the laser tracks geometrical characteristics (width, height, area, penetration depth) and the processing parameters are established. The proposed analytical relationships look like $y = a_0\left(P^{\alpha_1} Q_m^{\alpha_2} V^{\alpha_3}\right) + b_0$ where y is one of these geometrical characteristics.

Two kinds of models are explored to predict the clad geometrical form and characteristics. The first one is an analytical model in which the powder distribution in the feed jet is supposed to govern the laser clad geometry. It is then proved that the powder distribution in the jet can't determinate the final clad geometry. In the second one the general form of the clad cross section is supposed to be a disk due to the surface tension forces. Analytical relationships are established between the radius and the disk centre on one hand and the process parameters on the other hand. Comparisons between experimental observations and simulated geometries are very convincing.

Thermal study is carried out using the one source point method and the so called Green functions. The microstructure is experimentally observed, analyzed and explained with the thermal modelling through the solidification front velocity and the temperature gradients calculations.

NOMENCLATURE

P	Laser power (W)
Q_m	Powder mass flow (g/s)
V	Laser velocity (m/s)
H	Laser clad height (m)
W	Laser clad width (m)
H_f	Molten pool depth in the substrate (m)
S	Cross section area of the molten pool above the substrate (m^2)
S_f	Cross section area of the molten pool in the substrate (m^2)
R	Virtual disk radius (m)
P_e	Powder efficiency
ρ	Density (kg/m^3)
Q_l	Absorbed laser energy in unit time [J/s]
Q_c	Heat flow crossing the molten zone toward the substrate [J/s]
Abs	Absorption
$G(x,y,z)$	Green function
$T(x,y,z)$	Temperature
T_m	Molten temperature (K)
T_0	Initial Temperature (K)
t	Time(s)
α	Diffusivity (m^2/s)
d	Diameter of the spherical molten zone (m)
k	Conductivity [W/(m.K)]
A	Interface area between the molten pool and the substrate (m^2)
r_p	Powder jet stream Radius [m]
L	Distance between the nozzle and the substrate surface [m]
l	Distance from the nozzle and the powder jet focus plan [m]
$a_0,b_0,\alpha_1,\alpha_2,\alpha_3$	Regression parameters

I. INTRODUCTION

I.1. Start with Laser Manufacturing

Since the discovery of the laser (Maiman, 1960), scientists have continued to develop its applications. The laser is playing an increasingly important role in our daily lives and in the industrial world. The progress made on the quality of lasers, their power and reliability, range of applications continues to grow in the industrial field. It extends from the medical field with eye surgery or skin treatment laser until the industry with the marking, cutting and welding.

Direct laser fabrication is a field of application of high power lasers issue from rapid prototyping. One or more materials, in powder form, are injected into a laser beam, to be melted and deposit after solidification to form a clad. Layer upon layer, it is possible to manufacture parts mono-or multi-material pieces. Metal pieces in 3D complex form can be realized. This new way of making objects, called additive manufacturing, could, according to some, become the preferred manufacturing technique and unavoidable in the future as it frees the imagination of designers, eases logistics and new achievements cannot be achieved using the conventional means. Already established in the automotive and aerospace manufacturing methods additive conquer space, art and jewellery. Meanwhile, many scientific studies are conducted worldwide to optimize the performance of this new process, especially the geometric accuracy and surface quality of parts produced. The main advantage of the technique is the absence of moulds and tools in the parts development, and the ability to quickly produce prototypes or functional parts of complex geometries.

The idea of introducing the laser in the world of manufacturing began with William M. Steen and V. M. Weerasinghe (Weerasinghe et Steen, 1983). The first work has been done at Sandia National laboratory (Light Engineered Net Shaping (LENS)) and at Los Alamos Laboratory (Direct Light Fabrication (DLF)) by (Mazumder et Singh, 1986), (Mazumder et Kar, 1987). Thin layers of metal are melted and deposit successively. They showed that a 3D object can be achieved directly from a database type CAD with properties equivalent or better to those obtained with other methods. In the following years several studies have started in the manufacturing of direct laser deposition based on laser cladding technique. In 1987 C.F. Marsden studied the absorption of the laser power during the deposition of a coating (Marsden et al., 1987), and in 1990, he studied the interaction of the powder with the laser (Marsden and al., 1990). In 1991, (Hoadley et al., 1991) describe by physical models, the individual influence of each process parameter on the clad characterization. In 1992 (Hoadley et Rappaz, 1992) developed a 2D model of the transfer heat to aid the powder mixes quickly and uniformly in the molten zone.

Technologies issues from rapid prototyping provide innovative solutions for the production of parts (reducing the time and new features). Rapid production from laser fusion of metal powder targets a market of complex shape, small size, high value and produced in small series. The market for additive fabrication becomes important and is currently growing: 24.1% increase in revenue related to the additive manufacturing in 2010 according to Wohlers report 2011.

I.2. Direct Laser Fabrication Process with Powder Projection

Direct Laser Fabrication (DLF) is a method for manufacturing parts with single or multi-materials directly from powders injected into the beam and without the presence of a binder. As shown in Figure 1, the powder melts and is deposited on the workpiece. The formed clad height varies from 0.1 to 2 mm, depending on the laser power, the material type and the amount of powder sent per unit length. The accumulation layer by layer allows the manufacture of the part.

The manufacturing machine used consists of a coaxial nozzle in relative motion with respect to a moving table on which is placed the substrate to initiate the manufacture of the piece. This machine can perform movements following five degrees of freedom for the realization of 3D parts with complex geometries that are not achievable without joining by conventional methods. Control of the process is made from CAD data. As shown in Figure 2, the focusing plane of the laser coincides with the substrate surface.

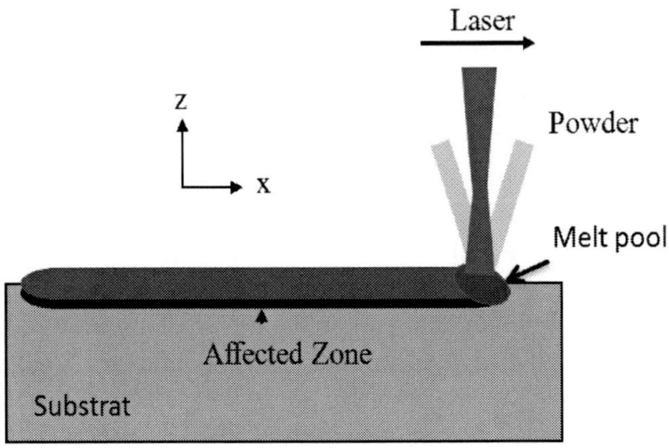

Figure 1. Schematic of (DLF) process by direct injection of powder in the laser beam.

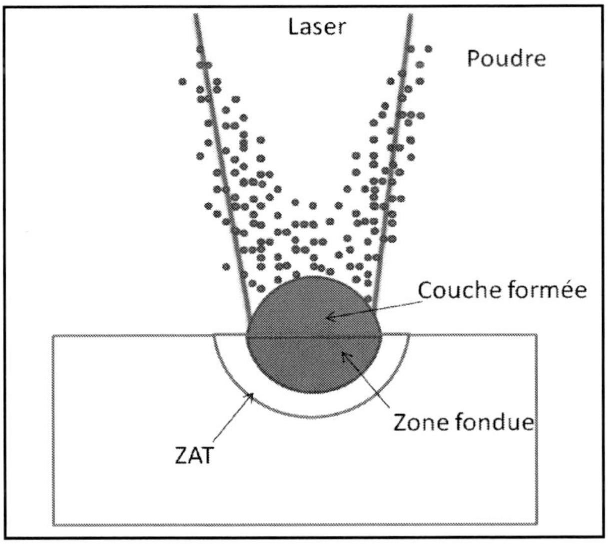

Figure 2. Principle of Direct Laser Fabrication (DLF).

The injected powder and part of the substrate melt and mix following the Marangoni effect movements in the melting pool. Once the first deposit is complete, the nozzle goes up a distance ∂z, defined in the CAD file, to form the second layer and so on. The velocity of the nozzle (powder + laser) allows the solidification of the melt and the formation of a deposit.

The different parameters which can influence the obtained clad are:

- The geometry of the powder jet stream (nozzle geometry, carrier gas flow rate, and forming gas flow rate);
- The position of the focusing plane of the powder jet stream;
- The powder mass flow;

- The laser beam divergence;
- The position of the of the laser focusing plane;
- The laser power and its distribution in the beam;
- The nozzle movement with regard to the building part.

It is necessary to analyze the influence of each parameter, and optimize experimental conditions to obtain the best possible part. In this study, the geometry of the powder jet stream, the focusing planes position, the laser beam divergence are fixed and supposed to be yet optimized. Influence of the three main parameters which are the laser power P, the velocity of the laser beam with regard to the building part V and the powder mass flow Q_m are studied.

I.3. Advantages and Weaknesses of the Direct Laser Fabrication Process

Compared to rapid prototyping processes, direct manufacturing obviously allows creating directly a functional part that can be tested or even used immediately in its environment with the actual loads. (DLF) offers several advantages over other conventional manufacturing methods:

Time Reduction

Under the pressure of international markets, time is a very important criterion for achievement of technical and commercial objectives. The (DLF) can meet this requirement because no moulds and tools are necessary and the transcription from the CAD model to the produced part is direct. It allows gaining competitiveness and responsiveness.

Flexibility

The (DLF) enables creation of parts with very complex geometries and without assembly. The design is free and so parts are more esthetical and lighter. The multi-material structures can further increase functionality and lightness. It is also possible to produce specific coatings of excellent quality, lower distortion and also to repair damaged parts with high adhesion (Toyserkani et al., 2005). This process is ideal for all productions of custom objects (mass customization).

Environment

Single pieces manufacturing from powders allows limiting transports of different elements and reducing or even removing machining procedures leading to the loss of material. This process is often described as favourable for environmental protection. Actually a serious assessment of its environmental impact remains to be done because the laser efficiency still low and powder has to be produced.

Thermal Control

The process uses a laser high energy to melt the under layer and the injected powder. The control of the melting zone geometry is possible through the input parameters, in particular laser power and velocity. The geometry of this melting zone must be optimized to improve the microstructure and minimize internal stresses because they determine the mechanical properties of the fabricated part.

II. Experimental Conditions

A 5 axes machine associated with a 700 W fiber laser (Ytterbium Fiber Laser System, diameter: 150 µm diameter, wave length about 1070-1080 µm) and a coaxial nozzle specially designed according to the laser beam characteristics is used in this study. A carrier gas (Ar) flow, fixed at 3L/min, import the powder while a secondary gas (Ar) flow, fixed at 5L/min, is shaping the powder stream. This nozzle is also equipped with a cooling channel to dissipate heat.

Figure 3 shows the conic powder jet stream geometry under the nozzle. Horizontal and vertical image analyses are studied. On the upper part of the jet, the horizontal image analysis displays two distributions due to the coaxial form of the nozzle. In the powder focus plan and under it only one distribution remains. An image analysis software is used to study the powder distribution in the jet from the light reflection. The obtained function is not directly the powder distribution in the jet but rather a projection of this distribution in the image plan. Nevertheless, it can be deduced that, in the focus plan, the powder distribution is an almost symmetrical Gaussian type distribution.

Multi-material

One of the opportunity offered by this technique, and therefore one of the issues to consider, is the realization of multi-material parts. Various kinds of powders can be injected simultaneously. Controlling their mass percentages, functionality of the part can be adjusted to the local solicitations.

Note also that some limitations exist for Direct Laser Fabrication process. It can't allow making parts with materials that crack at hot or cold temperature or which are too viscous at liquid state. In addition, a post-processing is currently required to remove the mounting supports and/or polish the part surface.

The main applications are:

- Aerospace and automotive (F1). Examples: TA6V, Inconel 718 (aircraft engines), aluminum alloys 2000, 7000, 6000 (fuselage).
- Medical field with biocompatible alloys. Examples: Stainless steels (904, 316), Cr-Co, TA6V.
- Tools or injection moulds. Examples: Steel 100Cr6, 42CD4, nickel base alloys.
- Art: jewellery, sculptures. Examples: Stainless Steel.
- Maintenance of weapons with damage parts repairing or even building of new parts no longer available on the market.

The vertical line in Figure 3a displays the distribution along the z axis. As shown in the image analysis, the higher concentration of the powder stream is in the focus plan. The coaxial nozzle allows a perfect symmetry of the particles distribution in the stream and very few particles outside the main trajectories. The powder focus plan, at a distance l (l=3.5 mm) from the nozzle, is a disk with a diameter d_p (d_p= 2 r_p = 0.6 mm). The work distance L is constant and equal to 5 mm. Under the powder focus plan the powder distribution remains a Gaussian type distribution. The distance between the focus plan and the substrate surface is (1.5 mm) and in the interaction plan the powder stream is 2 mm in diameter. The laser beam is focused on the substrate surface while the focused plan for the powder stream is 5 mm above. Figure 4 shows the uniform laser power distribution in the focus plan. The diameter is 0.53 mm.

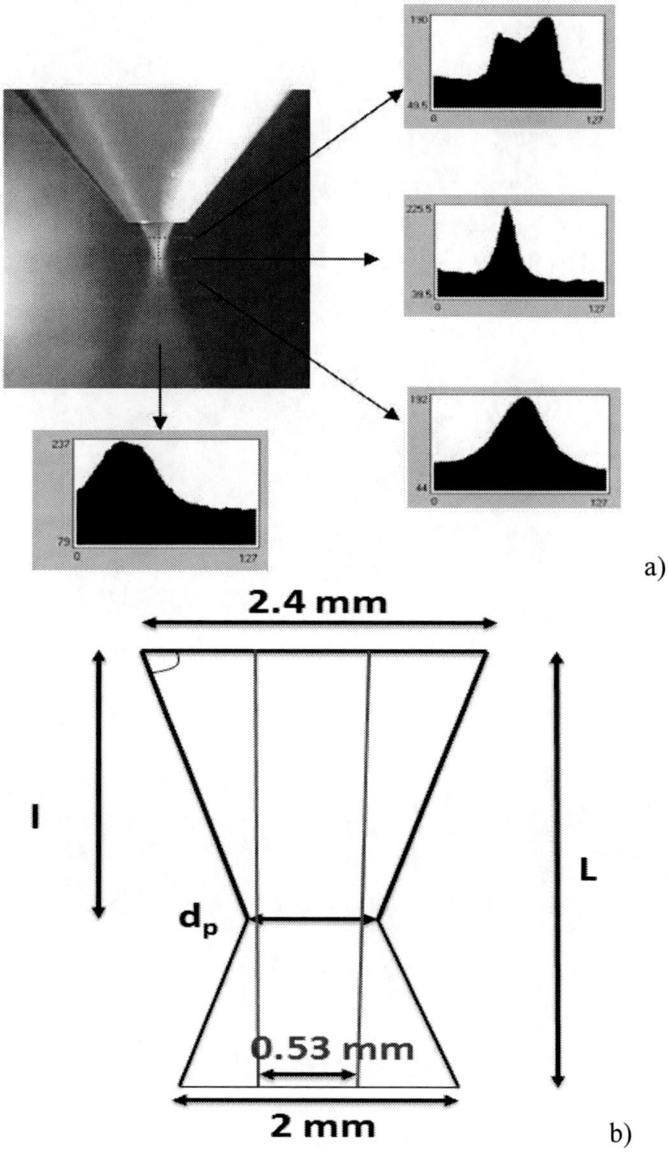

Figure 3. (a) Powder stream pictures analysis. (b) Powder cone shape from the nozzle issue to the laser focused plan.

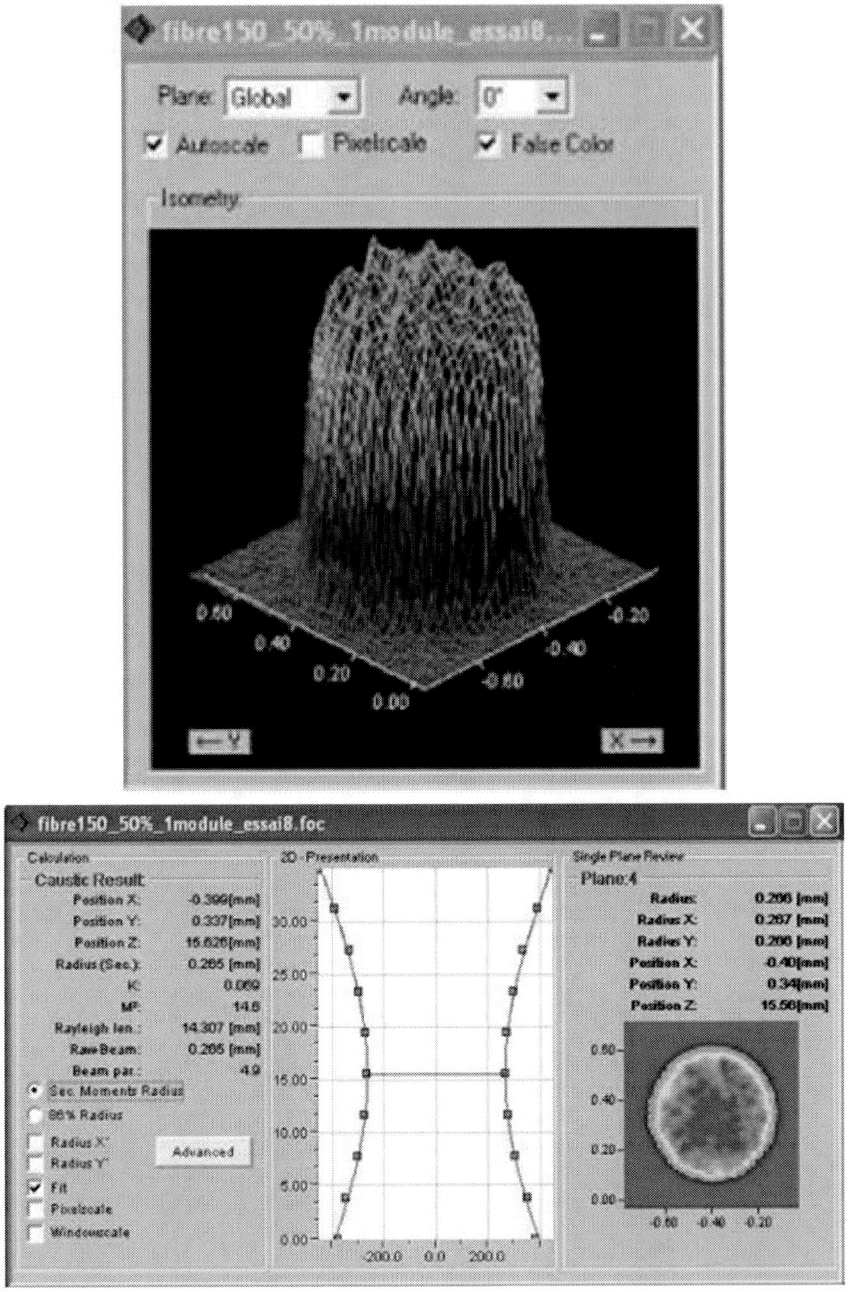

Figure 4. Laser distribution at the interaction with the substrate (plan of focalization).

In this study, 316L with particles size between 45 and 90 μm are used to build the experimental single laser tracks on the steel substrate (Figure 5). For each single laser track the laser power, the powder feed rate and the relative velocity are kept constant. The process duration is adjusted in order to obtain in each case the same length track of 90 mm. Each single laser track is produced on the same steel plate with same dimensions shown in Figure 6. To avoid any board effect, laser tracks are carried out at a distance of 10 mm from the substrate boards.

Figure 5. 316L Powder.

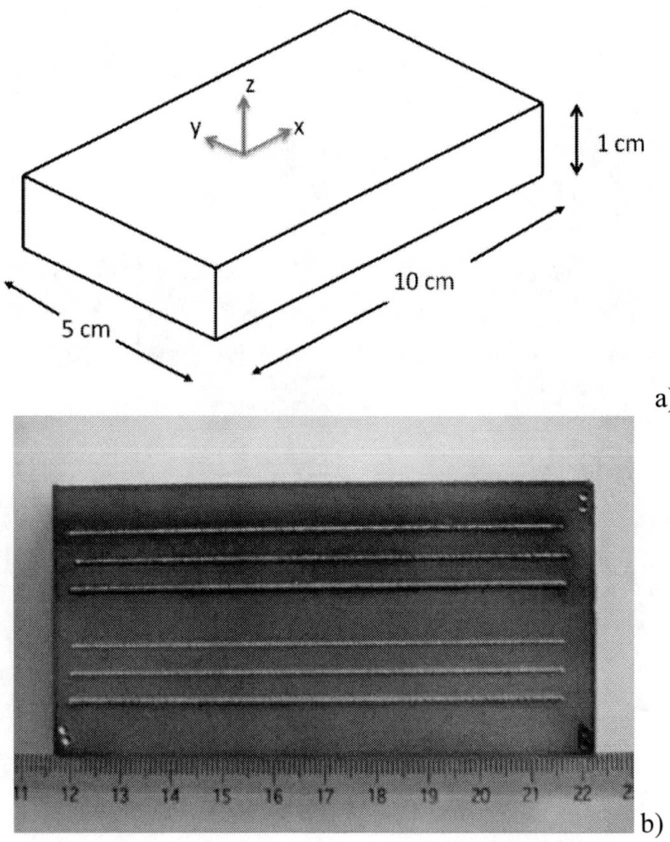

Figure 6. a) Substrate dimensions. b) Single tracks on the steel substrate.

Experimental results are analysed with observations on laser tracks cross sections. Cross section characteristics change with power, velocity and powder feed rate. Characteristics of each single track are evaluated measuring the clad height H, the clad width W, the penetration depth H_f, and areas S and S_f. All these dimensions are presented in Figure 7.

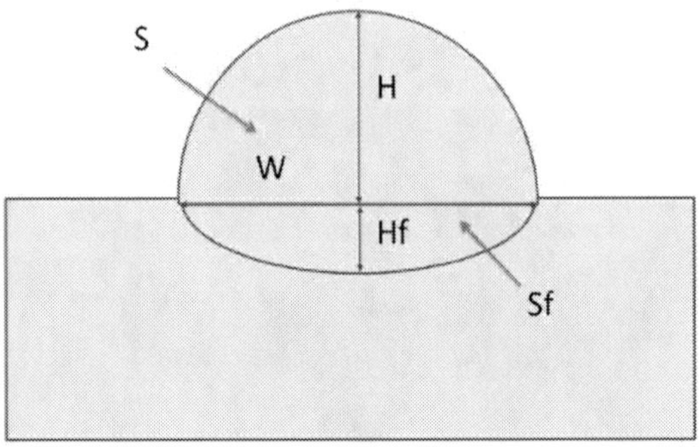

Figure 7. Cross section characteristics of a single laser track.

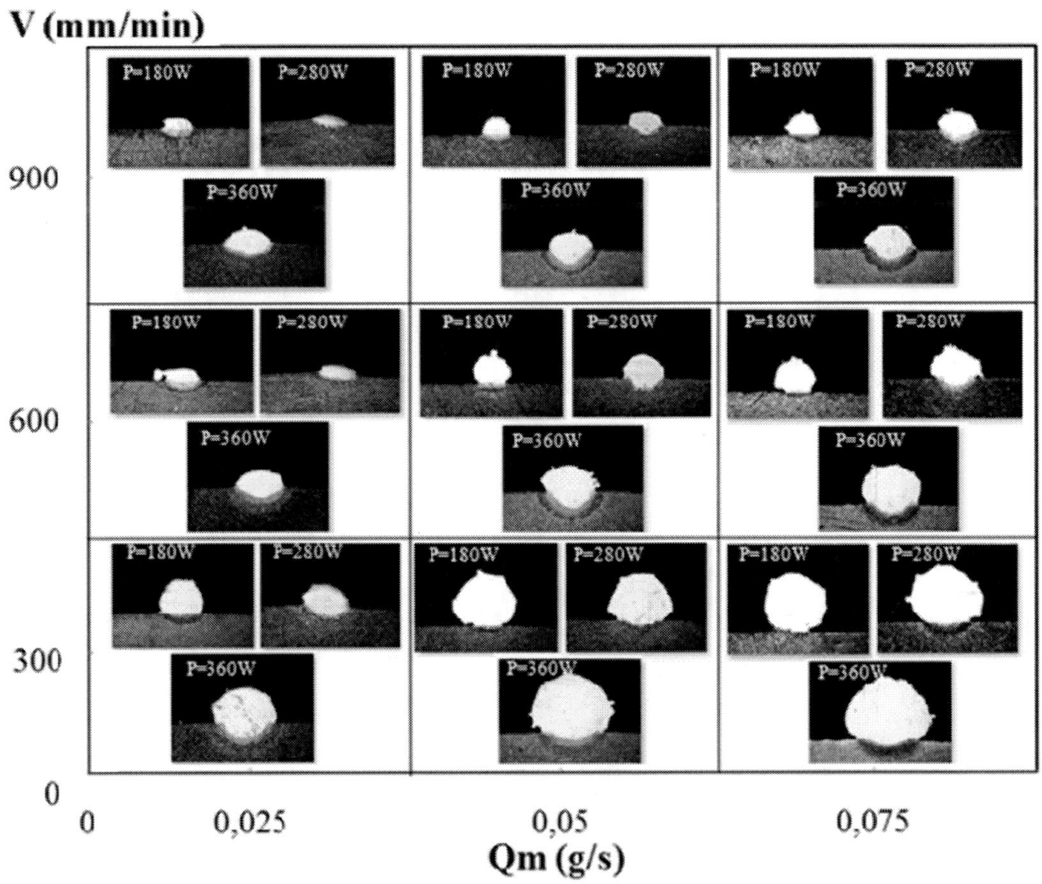

Figure 8. Single laser tracks cross-sections.

To study the effects of the main laser cladding parameters on the clad geometry, tracks were produced in a wide variety of power P (180, 280 and 360 W), velocity V (300, 600 and 900 mm/min) and mass feed rate Q_m (0.025, 0.05 and 0.075 g/s). Samples are polished and etched with Nital (2%). Every combinations of the 3 values of the 3 parameters are explored which represents 27 single tracks as shown in Figure 8. For each one 2 cross-sections are studied and the mean result is reported in (Table 1).

Table 1. Experimental measures for each single laser track

	P	Q_m	V	W	H	S	H_f	S_f
Clad N°	**W**	**g/s**	**mm/min**	**mm**	**mm**	**mm²**	**mm**	**mm²**
1	180	0,025	300	0,56	0,47	0,25	0,03	0,01
2	180	0,025	600	0,50	0,19	0,07	0,05	0,01
3	180	0,025	900	0,43	0,15	0,05	0,05	0,08
4	180	0,050	300	0,41	0,73	0,53	0,00	0,00
5	180	0,050	600	0,38	0,39	0,19	0,00	0,00
6	180	0,050	900	0,42	0,24	0,09	0,02	0,00
7	180	0,075	300	0,28	0,81	0,59	0,00	0,00
8	180	0,075	600	0,39	0,47	0,22	0,00	0,00
9	180	0,075	900	0,40	0,33	0,12	0,00	0,00
10	360	0,025	300	0,83	0,49	0,36	0,22	0,11
11	360	0,025	600	0,72	0,27	0,15	0,15	0,07
12	360	0,025	900	0,68	0,20	0,09	0,15	0,07
13	360	0,050	300	0,86	0,82	0,76	0,13	0,06
14	360	0,050	600	0,71	0,42	0,27	0,21	0,09
15	360	0,050	900	0,62	0,26	0,14	0,22	0,08
16	360	0,075	300	0,90	0,90	0,89	0,07	0,03
17	360	0,075	600	0,73	0,59	0,41	0,14	0,06
18	360	0,075	900	0,71	0,35	0,18	0,17	0,07
19	280	0,025	300	0,66	0,32	0,17	0,15	0,07
20	280	0,025	600	0,60	0,14	0,06	0,07	0,04
21	280	0,025	900	0,53	0,09	0,03	0,07	0,03
22	280	0,050	300	0,64	0,68	0,52	0,06	0,03
23	280	0,050	600	0,62	0,37	0,19	0,15	0,05
24	280	0,050	900	0,55	0,22	0,09	0,11	0,04
25	280	0,075	300	0,65	0,80	0,70	0,03	0,01
26	280	0,075	600	0,56	0,47	0,28	0,12	0,04
27	280	0,075	900	0,56	0,27	0,12	0,13	0,04

III. Effects of Process Parameters

III.1. Effects on the Depth Penetration H_f

For the lowest power (180 W), fusion of the substrate still low and is even nonexistent in half cases (Figure 9). The molten zone reaches about 50 microns in cases 2 and 3, only 20 microns in cases 1 and 6 and is not visible in other cases. This induces a high relative uncertainty on these measures. However, in all cases, a deposit is actually performed and adheres to the substrate. So, a part of the powder and the substrate melt at least superficially.

In low laser power conditions and for constant speed, the molten zone depth decreases when the powder mass flow increases.

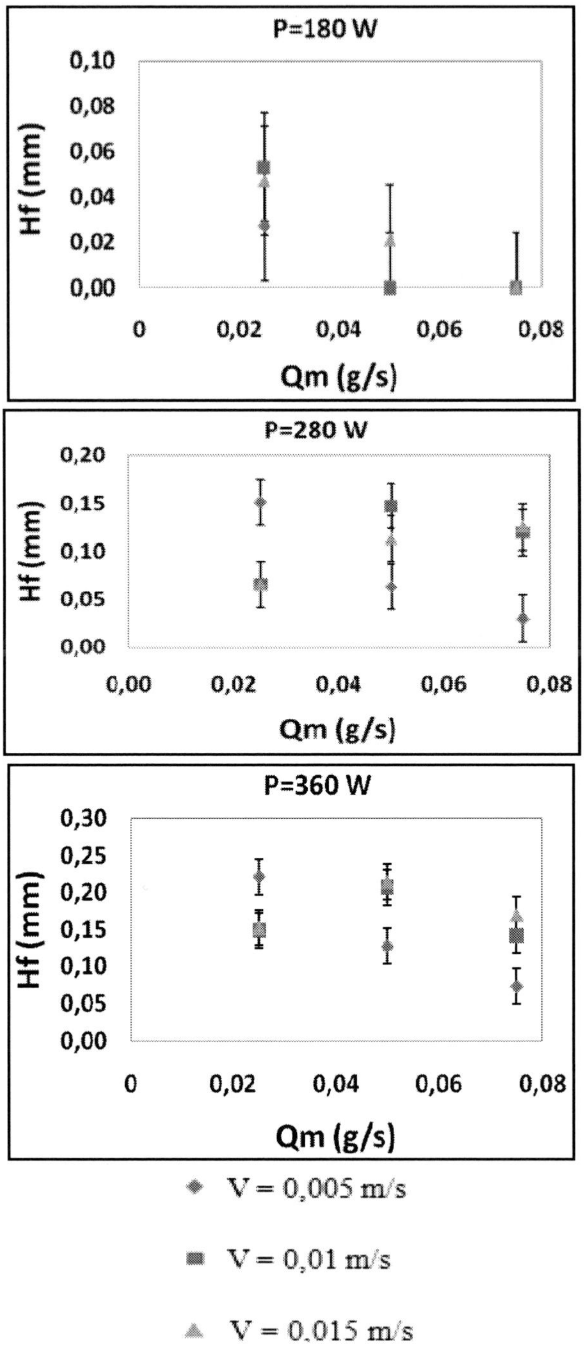

Figure 9. For each laser power, observations of the melt depth variation Hf as a function of mass flow Q_m for the three velocities V.

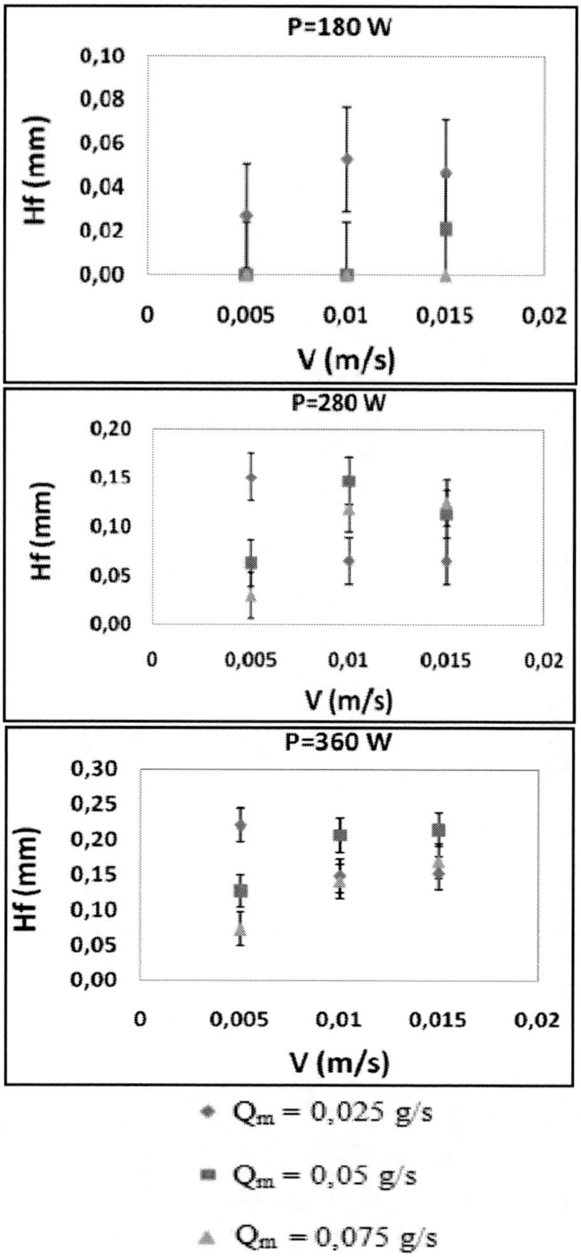

Figure 10. For each laser power, observations of the melt depth variation H_f as a function of velocities V for the three mass flows Q_m.

Two interpretations can be advanced at this level to explain this relationship. On one hand, the powder mass flow increase induces an increase in the number of powder particles between the nozzle and substrate. This increase of powder grains leads to a larger part of the laser beam absorbed in the powder jet stream. And consequently a lower part of the laser beam reaches the substrate. On the other hand, we notice that a mass flow increase implies a clad height increase and so the laser-clad interaction surface climbs away from the substrate, at least in part. These two phenomena, heat source masking and interaction plan shift, lead to a reduction of the molten zone depth when the mass flow increases. For laser power of 280 W and 360 W, the same effect is strictly noticed only for a speed of 0.005 m/s. For the two other speeds, the melt depth variation is not monotonous and reaches a maximum for the intermediate speed. This illustrates the interaction between the various process parameters and the complexity that can be encountered in predicting the geometry of the laser track.

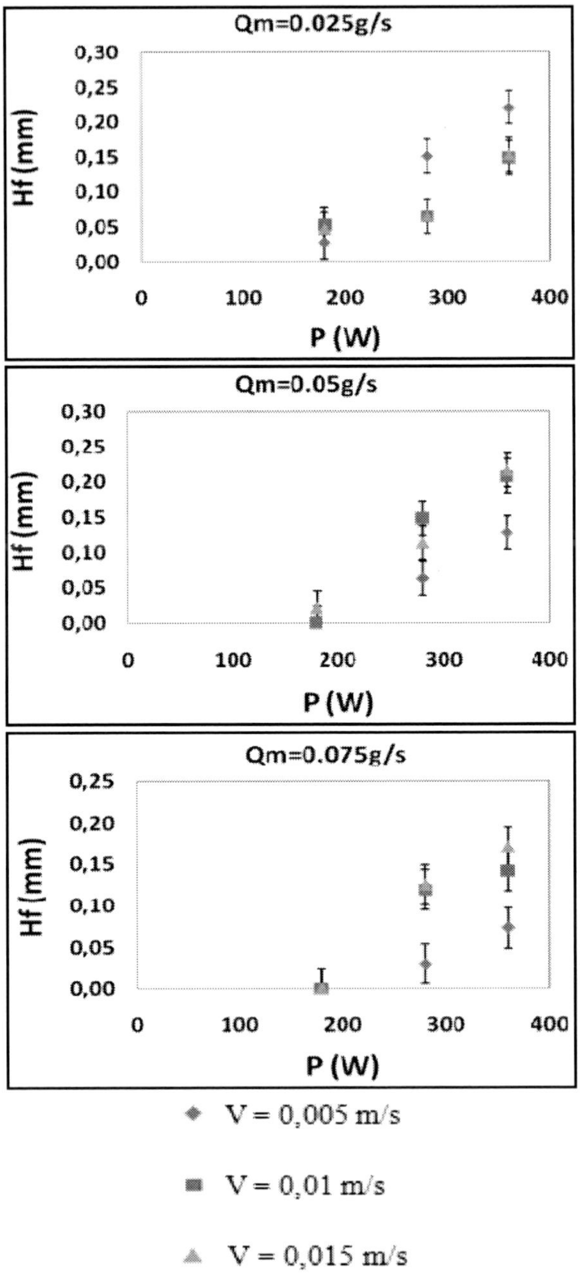

Figure 11. For each mass flow Q_m, observations of the melt depth variation H_f as a function of laser power P for the three velocities V.

Figure 10, does not meet either the expected variations. As speed increases the melt depth does not decrease systematically, as it could be expected due to the laser fluence decrease. This can be explained by the concurrency of this phenomenon with the interaction plan shift. As the speed increase the powder mass delivered by meter decreases, so the laser clad height decreases and the interaction surface (laser-clad) comes nearer the substrate which favours the substrate melting. Because of this concurrency, the melted zone depth is very difficult to predict. This makes more necessary the need to simulate the process.

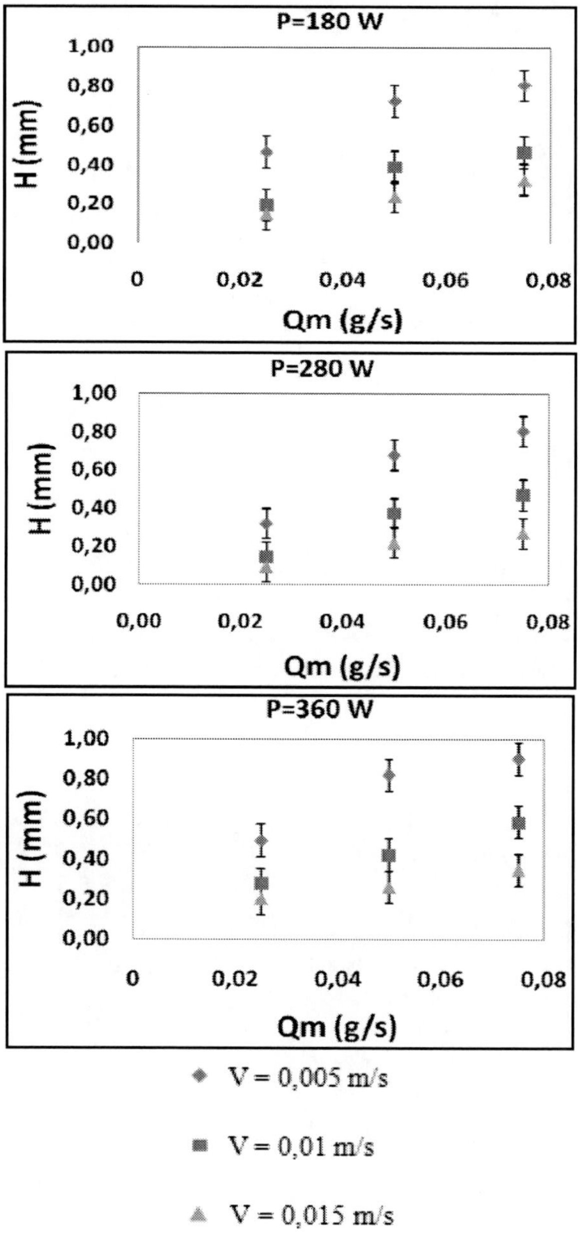

Figure 12. For each laser power, observations of the variation on the height of the deposit H as a function of mass flow Q_m for the three velocities V.

The process analysis should not be limited to the influence of each parameter in isolation. Figure 11 clearly shows that when other parameters remain unchanged, the laser power increase induces an increase of the molten zone depth. This relationship is not linear. In conclusion, the laser power is the parameter that has the clearest effect on the melted zone depth. The nozzle speed and the powder mass flow modulate the laser power effect varying the distance between the interaction surface (laser-clad) and the substrate.

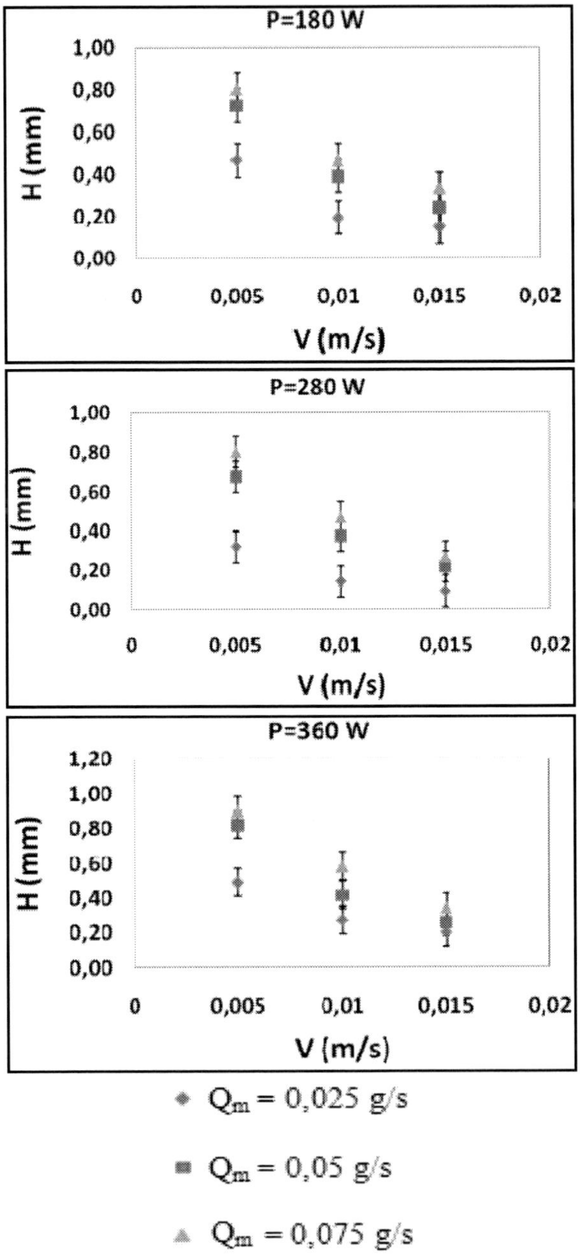

Figure 13. For each laser power, observations of the variation on the height of the deposit H as a function of velocities V for the three mass flows Q_m.

III.2. Effects on the Clad Height H

Figure 12 shows the powder mass flow and the laser power effects on the deposit height. For a constant laser power and speed, the deposit height increases with the powder mass flow. For the lowest speed 0.005 m/s (300 mm/min), the deposit height is the highest. As it is intuitively expected, the height H increases with the mass flow increase and a light saturation can be observed in most cases for higher powder mass flow.

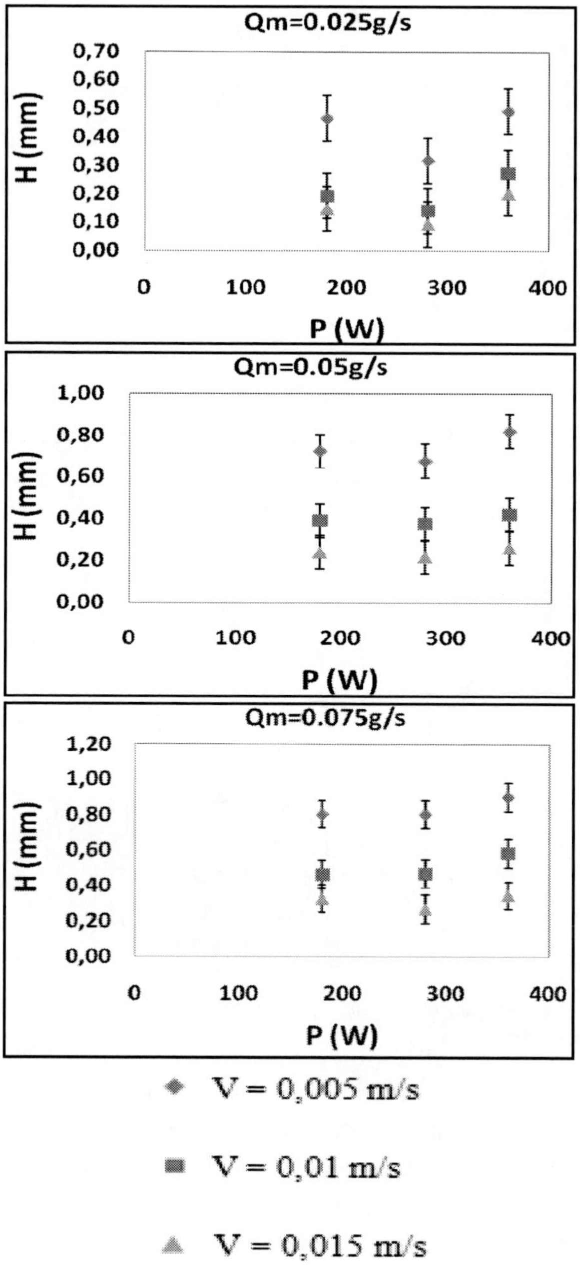

Figure 14. For each mass flow Q_m, observations of the variation on the height of the deposit H as a function of laser power P for the three velocities V.

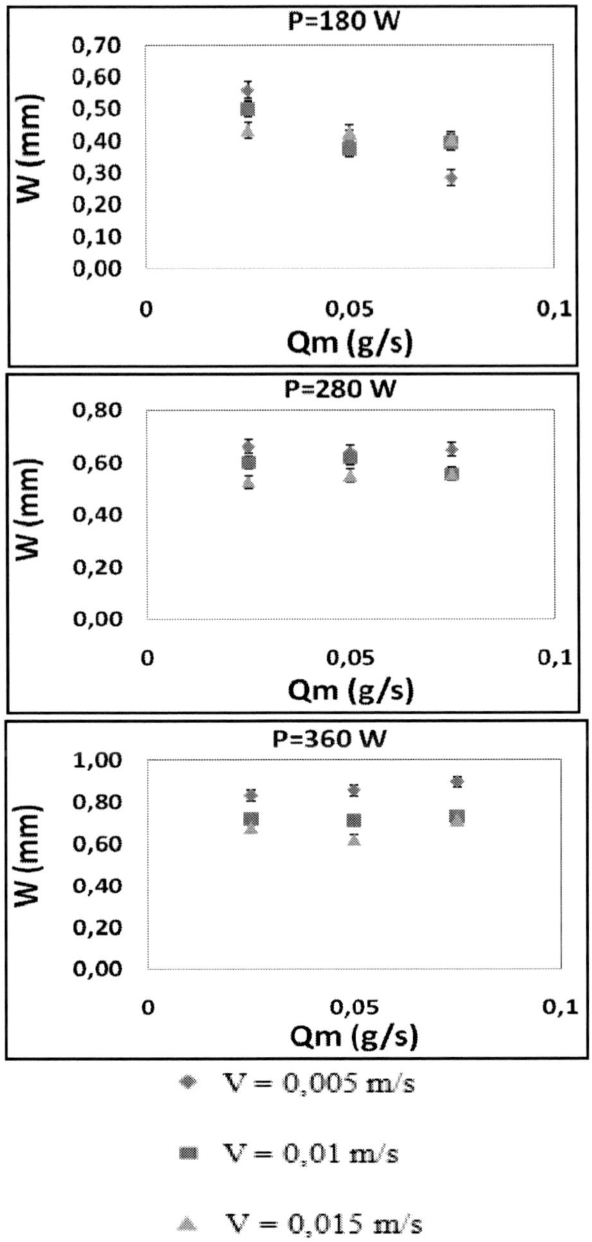

Figure 15. For each laser power, observations of the variation on the width W as a function of the mass flow Q_m for the three velocities V.

Figure 13 shows the laser speed and power effects on the deposit height for the three mass flows. For constant laser power and powder mass flow, the deposit height decreases with the laser speed increase. For the lowest mass flow (0.025 g/s), the deposit height is also the lowest. It can be considered that with the speed increase, the height decreases in a hyperbolic way. This was expected because the deposit height is almost proportional to the deposit mass per meter and so proportional to the ratio Q_m/V. Figure 14 shows that the laser power has few effects on the clad height. Curves are almost horizontal. The laser speed and powder mass flow have a major influence on the deposit height.

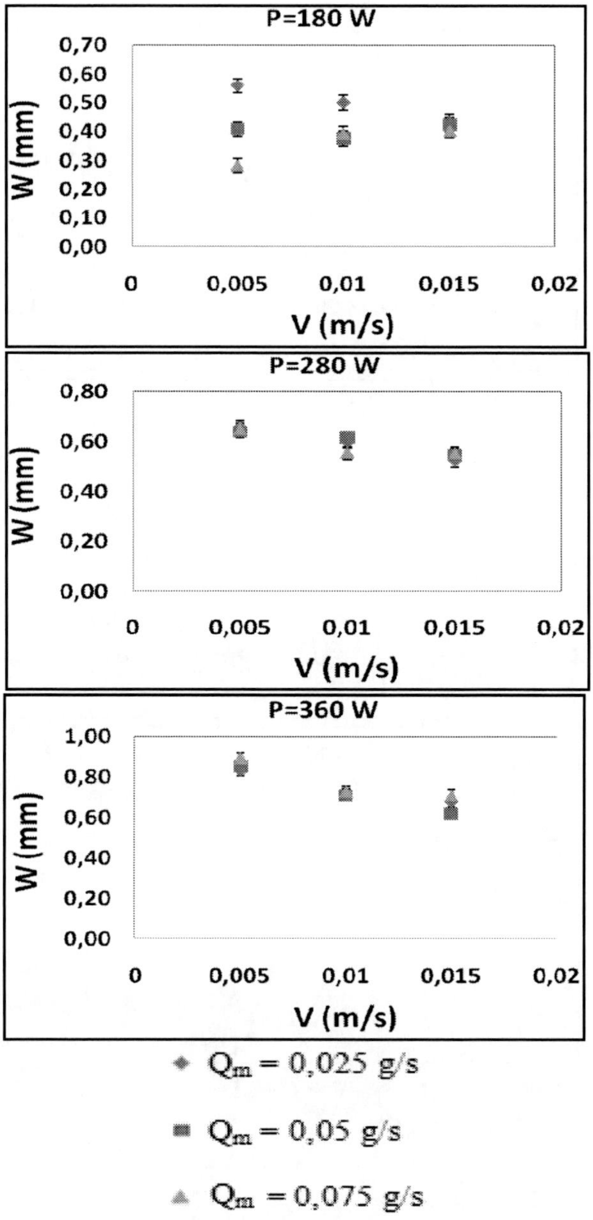

Figure 16. For each laser power, observations of the variation of the width W as a function of velocities V for the three mass flow Q_m.

III.3. Effects on the Clad Width W

The deposit width W is considered at the substrate surface and is equal to the molten zone width in the substrate. It does not always correspond to the maximum width of the deposit clad. In Figure 15, for a laser power of 180 W, width variations as the deposit mass flow increases are not significant. So, the powder mass flow has no influence on the molten zone width.

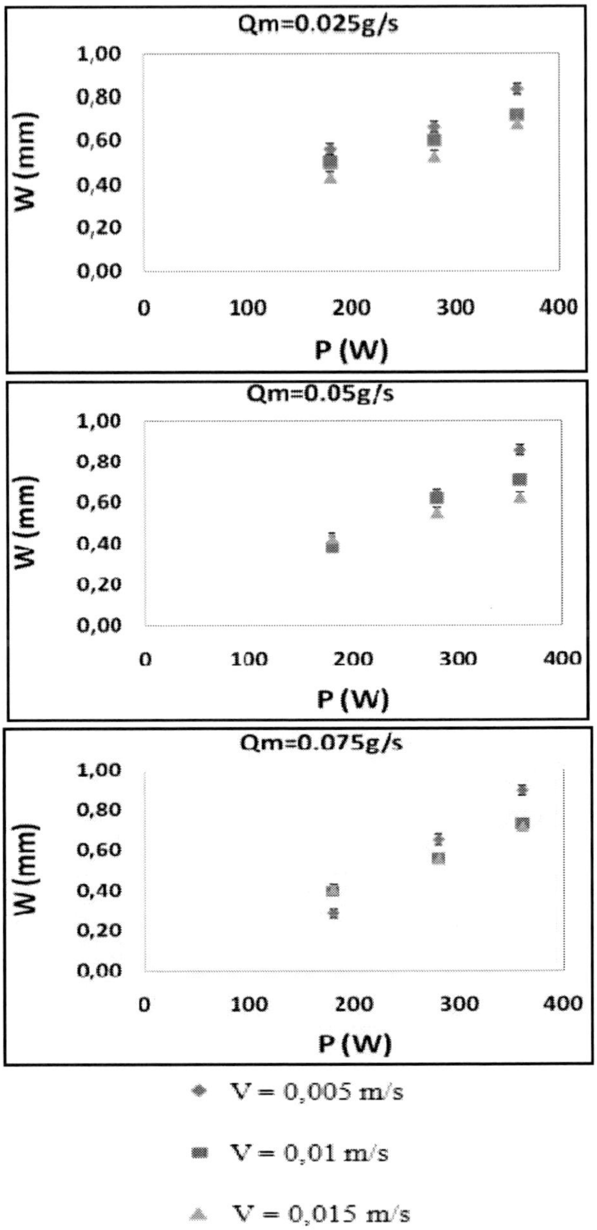

Figure 17. For each mass flow Q_m, observations of the variation of the width W as a function of laser power P for the three velocities V.

Figure 16 allows analyzing the clad width variations according to the laser speed. For laser powers of 280 and 360 W, the width decreases as the speed increases. This can be expected because the absorbed power per unit length decreases and so the molten zone volume decreases. The same effect is observed in the case of laser remelting. In the case of the lowest laser power (180 W), deposit width variations are more complicated to explain.

With the lowest powder mass flow (0.025 g/s), the deposit width decreases as the laser speed increases, whereas with the highest powder mass flow (0.075 g/s), the width increases as the speed increases. The competition between variations of the absorbed power per unit length and the distance laser-clad interaction surface can explain these apparently surprising results. The laser speed increases and the powder mass flow decreases leads to a smaller deposit clad and so the laser-clad interaction surface is closer the substrate. As the laser speed increases and the powder mass flow decreases lead to a smaller melting bath the laser-clad interaction surface subsiding lead to bigger melting bath at substrate surface level. This phenomenon effect on the melting bath width is the highest when the laser power is the lowest (180W). In this case and for the lowest powder mass flow, the absorbed power per unit length decreases is the dominant effect and for the highest powder mass flow the laser-clad interaction surface subsiding is dominant.

Figure 17 shows a simple relationship between the molten zone width and the laser power. The more the laser beam power is high the more the deposit width is large. Again the same variations as in the simple laser remelting case are observed.

IV. PARAMETRIC STUDY

Effects of the different parameters analysed, more quantitative relationships are searched between the clad geometry characteristics and the experimental parameters (P, Q_m, V). The clad mass is measured after each single track fabrication for the two power series (180 and 360). In the 280W serial case, the mass measure has not been possible for a technical problem.

Being capable to predict the laser track characteristics is really important to build a wall and finally to fabricate an object. (Onwubolu et al., 2007) chosen a regression analysis as $Y = \propto_0 + \propto x_1 + \propto_2\ x_2 + \ \ldots . \propto_n x_n$ where x_i are the process parameters. Unfortunately, correlation coefficients presented are quite low. U. De Oliveira (De Oliveira et al. 2005) searched and founded more persuasive relationships like $\left(P^{\alpha_1} Q_m^{\ \alpha_2} V^{\alpha_3}\right)$. In our study, we also made the assumption that some combinations as $\left(P^{\alpha_1} Q_m^{\ \alpha_2} V^{\alpha_3}\right)$ are able to predict the different laser track geometry characteristics. Using the Log function a linear regression is used to determine α, β and γ. We finally obtain relationships as $y = a_0\left(P^{\alpha_1} Q_m^{\ \alpha_2} V^{\alpha_3}\right) + b_0$ where y is one of the laser track geometry parameters while a and b are constants. With our experimental results, H is found proportional to $P^{1/4} Q_m^{\ 3/4} V^{-1}$ with a correlation coefficient R=0.97 (i.e. R^2=0.94) taking into account all the 27 experiments. (Figure 18) shows these results.

The powder mass which falls on the substrate per meter is $Q_m\ V^{-1}$ (This is also the relationship proposed by U. De Oliveira). However, the laser power plays a non negligible part because it melts powder grains and the more the amount of melted powder increases the more the clad height increases. The correct representation of the relationship between the clad height and the process parameters could be written in the form $P^{1/4} Q_m^{\ 3/4} V^{-1} = \left(\frac{P}{Q_m}\right)^{1/4} \frac{Q_m}{V}$. The first factor takes notice of the laser power effect on the powder melting. The R coefficient value of 0.97 shows a good correlation and the uniform distribution of the residuals confirms the linear model. The experimental repeatability uncertainty for the clad height is 0.040mm. If the linear regression formula is used to predict the clad height from the process parameters the uncertainty on this calculated clad height is 0.060mm. These two values are sufficiently narrow to show that this linear relationship between H and $P^{1/4} Q_m^{3/4} V^{-1}$ is relevant. (Figure 19).

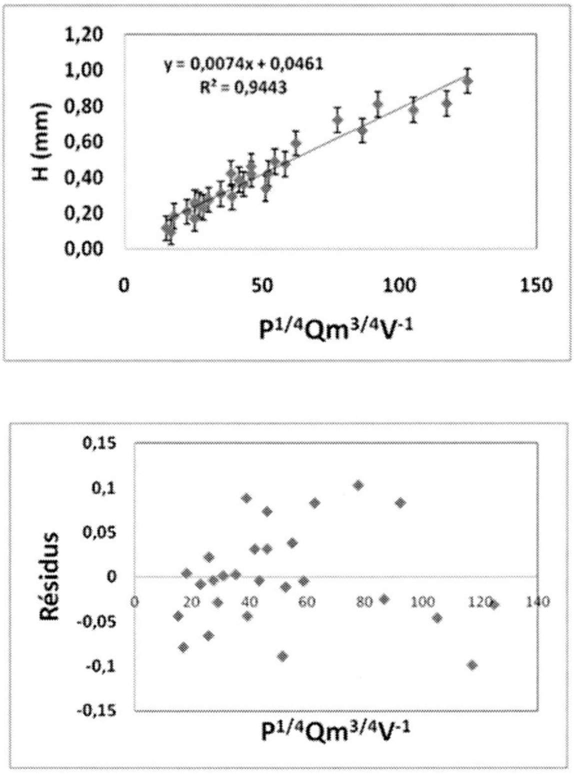

a) b)

Figure 18. a) Laser track height H versus $P^{1/4}$ $Qm^{3/4}$ $/V^{-1}$. b) Residuals.

The clad width is also proportional to a similar process parameters combination, with $\alpha_1 = 0.75, \alpha_2 = 0$ and $\alpha_3 = -0.25$. It is to be noticed that the mass feed rate Q_m has a negligible influence on the clad width. The two parameters laser power and velocity are sufficient to determine the single track width which could be the sign of a thermal effect like in the re-melting process. With our experimental measures a correlation coefficient R= 0.96 is found with $P^{3/4} V^{-1/4}$. The experimental repeatability uncertainty concerning the width is 0.040 mm, whereas uncertainty on the estimated width using the linear dependence between W and $P^{3/4} V^{-1/4}$ is 0.068mm which is comparable with the previous height uncertainty.

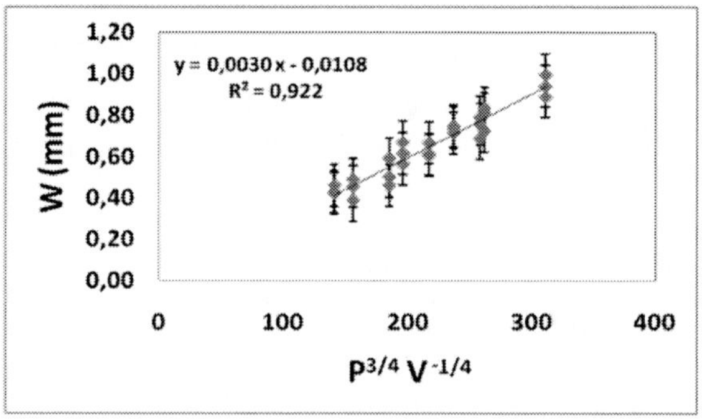

a)

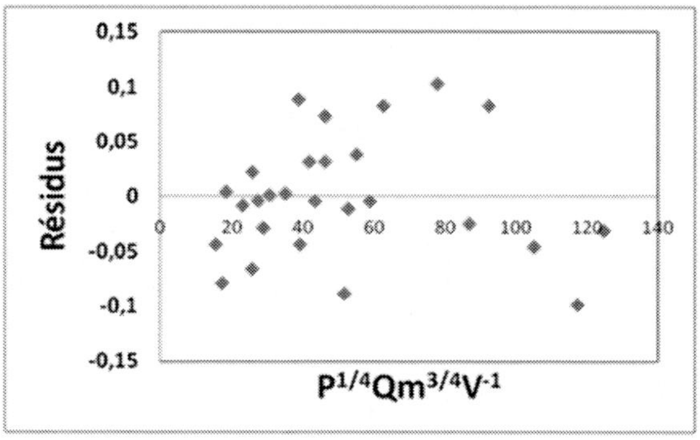

b)

Figure 19. a) Clad width versus $P^{3/4}/V^{1/4}$. b) Residuals.

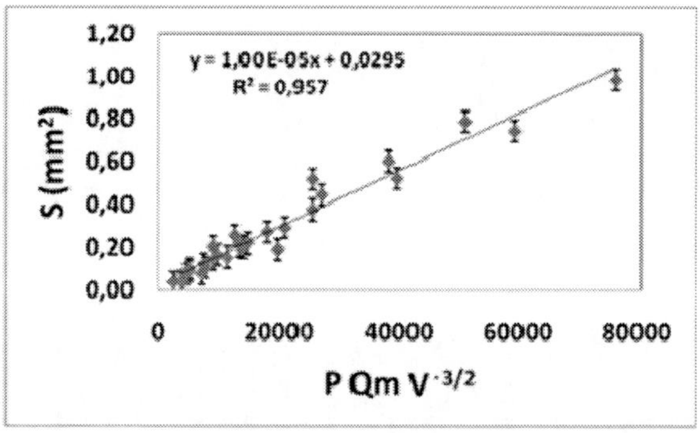

a)

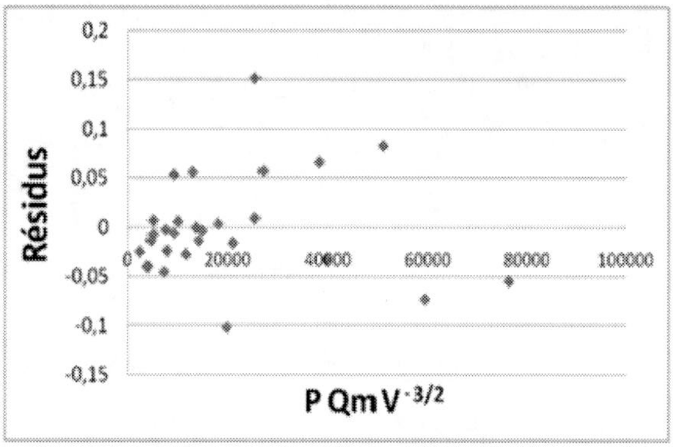

b)

Figure 20. a) Clad area versus P Q_m $V^{-3/2}$. b) Residuals.

Concerning the clad section area, the same method leads to the combination $PQ_mV^{-3/2}$ with a coefficient R equal to 0.978. The repeatability uncertainty is 0.025mm² and the uncertainty obtained for the estimated surface with the linear relationship is 0.049mm². (Figure 20)

Of course the clad mass is also proportional to $PQ_mV^{-3/2}$ as the clad mass is proportional to the cross section surface (Figure 21).

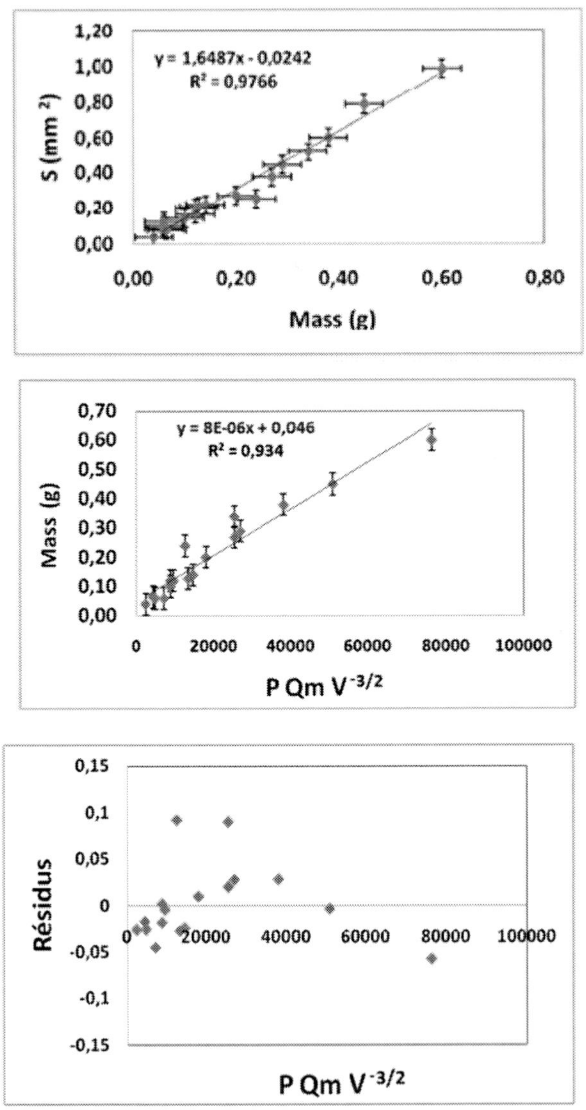

Figure 21. The linear dependence between the cross section surface and the clad mass.

The relationship between experimental parameters and the powder efficiency measuring the proportion of powder mass which stays on the substrate (Table 2) has also be determined. The uncertainty is 4.1% for the repeatability uncertainty and 5.2% for the estimated one. This is an important criterion because one of the Direct Laser Process goals is to do not lose matter by machining.

These relationships are very useful to predict the clad width and height but not the complete geometry and do not help to understand the physical phenomenon involved during the process. Another way to predict the clad geometry is to simulate the supposed physical phenomenon involved.

Table 2. Relationship between geometrical characterization and process parameters

	Relationship	Correlation coefficient (R^2)
H	$P^{\frac{1}{4}}Qm^{\frac{3}{4}}/V$	0.944
W	$P^{\frac{3}{4}}V^{\frac{1}{4}}$	0.922
S	$P\ Qm\ /V\sqrt{V}$	0.957
m	$P\ Qm\ /V\sqrt{V}$	0.934
$\mathbf{P_e}$	$P^{\frac{3}{4}}Q_m^{\frac{-1}{3}}V^{\frac{-1}{2}}$	0.910
$\mathbf{S_f}$	$Ln\ (P^{\frac{4}{5}}\ Q_m^{\frac{-1}{4}})$	0.649
$\mathbf{H_f}$	$Ln\ (P^2V^{\frac{1}{4}}\ Q_m^{\frac{-1}{4}})$	0.765

V. Prediction of the Geometrical Form of the Clad

The deposited layer results from powder accumulation. First the clad height is supposed to be proportional to the number of powder grains which fall on each point. The clad geometry is then directly induced by the powder distribution in the jet. Three powder distributions are tested (El Cheikh et al. 2012) in order to explore various possible cases.

The powder distribution is given by f (x, y) and expressed in units: $kg.m^{-2}.s^{-1}$. The mass which fell on a little surface (ds=dx.dy) after the passage of the laser is given by:

$$dm = \int_0^t f(x, y).dx.dy.dt$$

The time t during which the powder fall on this surface is given by:

$$t = \frac{b}{V}$$

where b is the chord of the powder deposition disk (Figure 22):

$$t = \frac{2\,\mathrm{r_p}}{\mathrm{V}}\sqrt{1 - \frac{\mathrm{y}^2}{\mathrm{r_p}^2}}$$

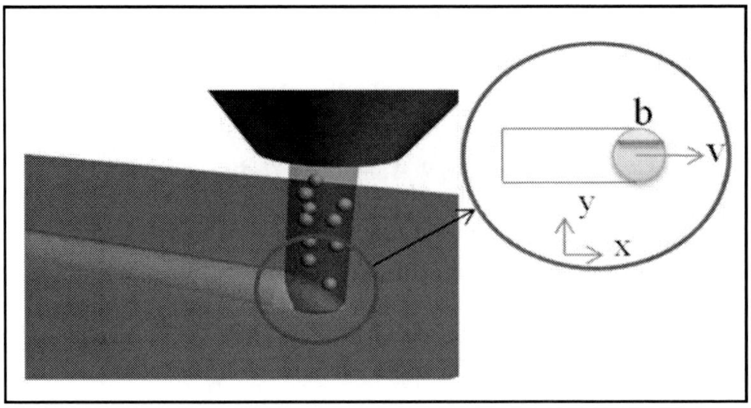

Figure 22. Powder deposition on the substrate surface.

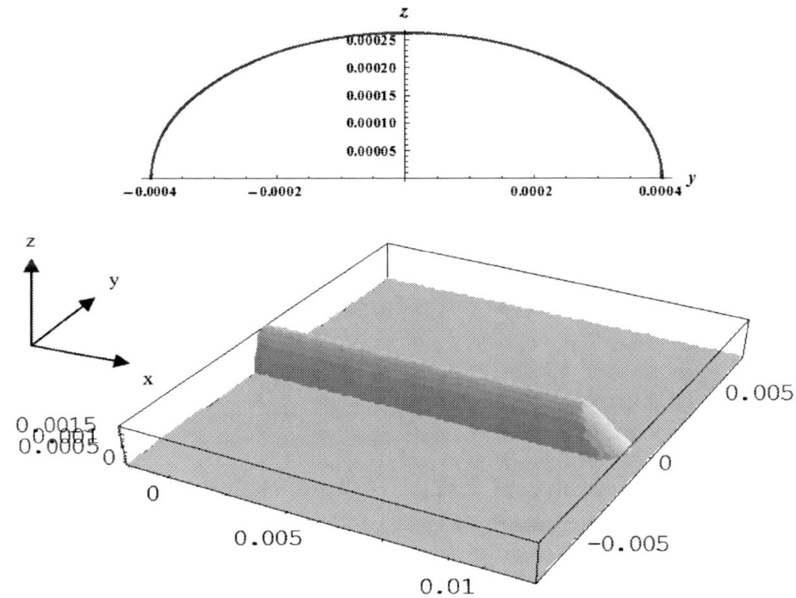

Figure 23. Geometry of the deposited layer in case of a uniform powder distribution.

The powder height h which fell under the studied surface behind the powder jet is given by:

$$\rho = \frac{dm}{h.dx.dy} \rightarrow h = \frac{dm}{\rho.dx.dy}$$

where ρ is the powder density.

And finally:

$$h = \frac{1}{\rho} \int_0^{\frac{2\,r_p}{V}\sqrt{1-\frac{y^2}{r_p{}^2}}} f\left(x,y\right).dt$$

Beneath the powder jet, b is only a part of the chord and so depends on x and y.

$$h = \frac{1}{\rho}\int_0^{\frac{b(x,y)}{V}} f(x,y) \,.dt$$

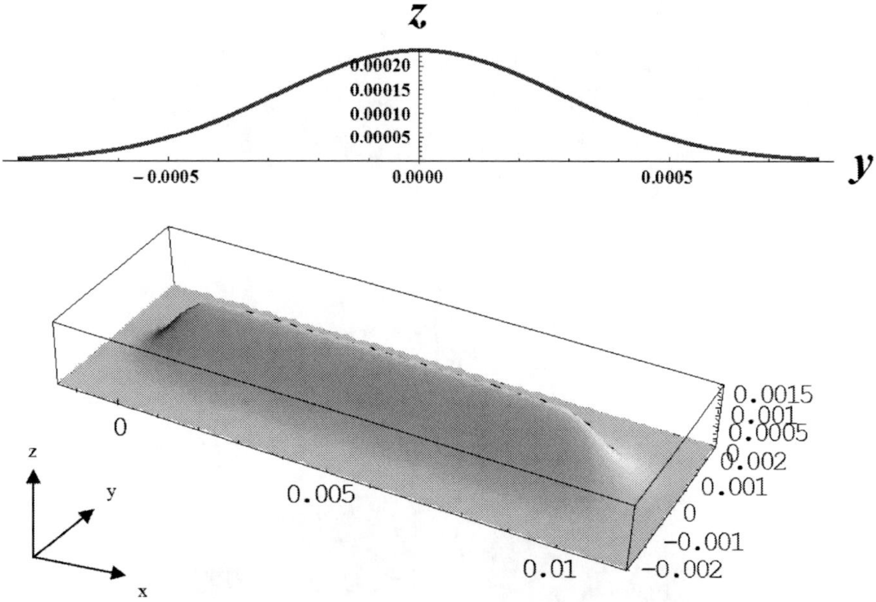

Figure 24. Geometry of the deposited layer in case of a Gaussian powder distribution.

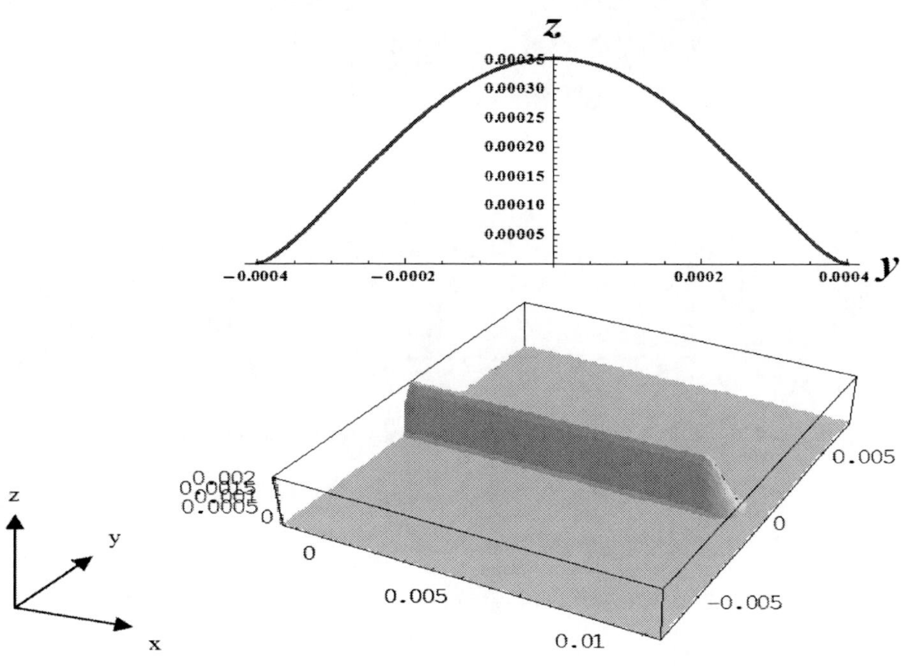

Figure 25. Geometry of the deposited layer in case of a polynomial distribution.

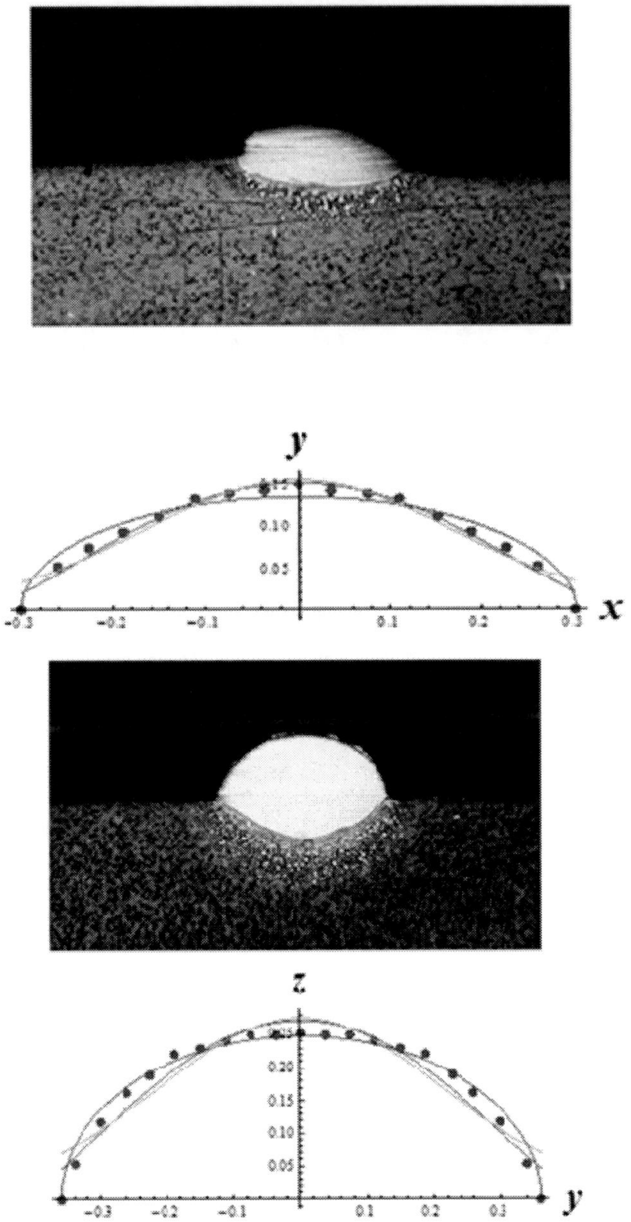

Figure 26. Fitting with three different functions to describe experimental results.

V.1. Uniform Distribution

If the powder distribution is uniform, the density of the powder feed rate is given by

$$f(x, y) = \frac{Q_m}{\pi r_p^2}$$

where Q_m is the powder feed rate and r_p the radius of the powder deposition circle.

One can find that, behind the powder jet, the profile is given by $\frac{2Q_m}{\rho\pi v r_p^{\,2}}\sqrt{r_p^{\,2}-y^2}$ and beneath the powder jet by

$$\frac{2Q_m}{\rho\pi v r_p^{\,2}}\sqrt{r_p^{\,2}-y^2}-(x-vt)$$

V.2. Gaussian Distribution

If the powder distribution is Gaussian along x and y axis:

$$f(x,y)=\frac{Q_m}{\pi r_p^2}e^{-\frac{(x-Vt)^2+y^2}{r_p^{\,2}}}$$

One can find that the profile behind and beneath the powder jet takes the following form:

$$\frac{2Q_m}{2\sqrt{\pi}\,v\,\rho.r_p^{\,2}}e^{\frac{-y^2}{r_p^{\,2}}}.[erf(\frac{x}{r_p})-erf(\frac{x-v.t}{r_p})]$$

V.3. Polynomial Distribution

If the distribution is supposed to be parabolic:

$$f(x,y)=ar^2+br+c=a(x+y)^2+b(x+y)+c$$

One can find that under the powder jet:

$$\frac{8Q_m}{3\pi\,v\,\rho.r_p^{\,4}}(r_p^{\,2}-y^2)^{\frac{3}{2}}$$

The equation behind the powder jet is

$$\frac{2Q_m[(x-v.t)^3+3(x-v.t).y^2-2y^2\sqrt{r_p^{\,2}-y^2}+r_p^{\,2}(-3(x-v.t)+2\sqrt{r_p^{\,2}-y^2})]}{\rho\pi v\, r_p^{\,2}}$$

and zero everywhere else.

One can notice that with this approach the laser power is not took into account. Concerning the height H, this is not really significant because H has been found experimentally proportional to $P^{1/4}Q_m^{3/4}V^{-1}$ and so in our case $P^{1/4}$ varies from 1 to 1.19 while $Q_m^{3/4}$ varies from 1 to 2.28 and V^{-1} from 1 to 3. On the other hand, it is critical for the determination of W which has been found proportional to $P^{3/4}V^{-1/4}$. This can be explained by the fact that the clad width is controlled by a thermal effect which depends directly on the power P. This parameter plays an important part on the powder melting in the jet and on the melting bath dimensions.

Figure 26 shows fittings on two representative laser tracks geometries, a spread one and a near semi-circular one. While the experimental cross section is represented by blue points, the blue curve corresponds to the uniform distribution, the red one to the polynomial distribution and the green one to the Gaussian distribution.

The Gaussian and polynomial distributions are more suitable for the spread geometry while the uniform distribution is better for the semi-circular one. One can use the proposed analytical relationships for specific conditions but it is not likely that the powder distribution in the jet changes from a Gaussian distribution to a uniform one as process parameters are modified. Furthermore in the case of the lower laser speeds the clad geometry is a part of a disk and none of these distributions can leads to such geometry. So, in our case, the laser clad geometry is not directly induced by the powder distribution in the jet. In order to simulate the laser clad process in the stationary case, it is more efficient to fix the cross-section geometry as a part of a disk.

V.4. Circular Form

V.4.1. Theory

Actually, we have experimentally observed that clad dimensions depend on process parameters while the cross section always remains a part of a virtual disk which centre can be below or above the work piece surface (Figure 8). This second modelling represents the clad cross-section by a circular form and the new request is to determine the centre location and the radius of this disk. The disk centre is found above the work piece surface when the velocity isn't great enough to avoid the accumulation of the deposited material. The usefulness of the proposed model is to predict the height, the width, the surface and the general form of the cross section in any cases.

This circular geometry suggests that surface tension forces are dominant. The calculation of the capillary length gives a result of 0.004 meter. The liquid bath has a characteristic length smaller than this capillary length and so the gravity effect is negligible and the liquid shape is almost spherical (circular in the cross-section plane and slightly lengthened along the laser displacement axis because of the laser movement).

A similar approach exist, using the Simpson's rule, but is limited to cases where the clad is flattened. It allows describing the clad height and width by Davis et al. (2006), to optimize clad parameters by Pinkerton and Li (2004) and to avoid excessive dilution and overlap porosity by Colaco et al. (1996).

This approach is more general and allows the prediction of the form, the height, the width and the area of the clad from the process parameters in all cases.

Relationships between the Geometrical Characteristics R, W and the Process Parameters

Figure 27 shows the virtual disk supposed to describe the laser track shape.

The disk is defined by its centre location and its radius R. They both depend on process parameters. W is the clad width and also the disk chord at z=0. Taking into account only the powder which is incorporated in the clad, the actual surface of the clad cross-section is given by $\frac{P_e Q_m}{\rho V}$.

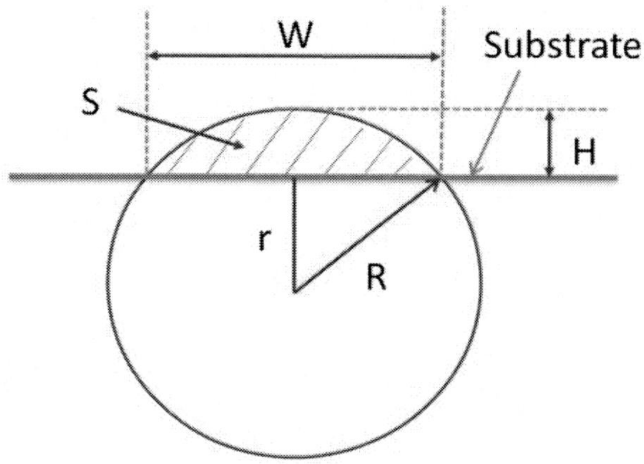

Figure 27. Schematic representation of the clad cross section.

Figure 28 shows the two kinds of clad which can be obtained during the deposition.

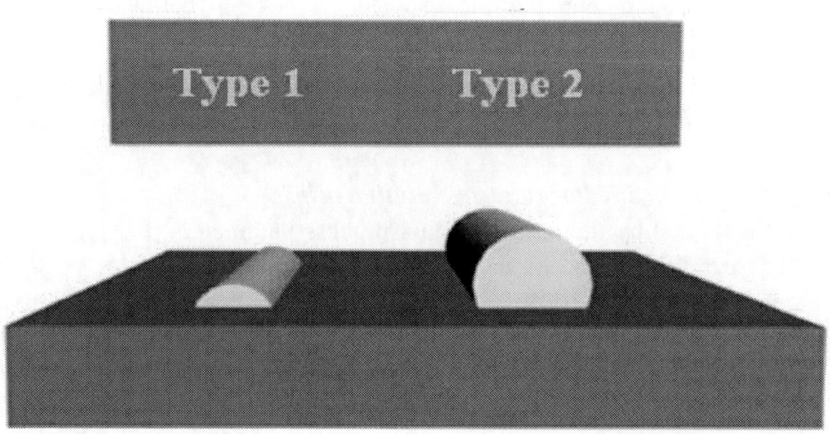

Figure 28. Observed kind of Clad.

They differ from their centre location which can be located below or above the substrate surface. The powder mass deposited is related to the laser clad volume above the substrate surface.

In the limit case the circle centre is located on the substrate surface and one can show that $\frac{P_e Q_m}{\rho V} = \frac{\pi R^2}{2}$. Then the disk radius is given by: $R = \sqrt{\frac{2\, Q_m\, P_e}{\pi\, \rho\, V}}$. This case is a very interesting one to build a structure because of its semi-circular geometry.

Outside this limit case:

- Type 1: If $\frac{P_e Q_m}{\rho V} < \frac{\pi W^2}{8}$, the disk centre is below the substrate surface and at a distance $\sqrt{R^2 - \left(\frac{W}{2}\right)^2}$ from it. The radius R is obtained by solving the equation:

$$\frac{P_e Q_m}{\rho V} = \frac{1}{6W}(R(1 - \sqrt{1 - (\frac{W}{2R})^2})(3R^2(1 - \sqrt{1 - (\frac{W}{2R})^2})^2 + 4W^2) \qquad \text{(V.4.1)}$$

This case is also interesting and corresponds to the flattened one.

- Type 2: If $\frac{P_e Q_m}{\rho V} > \frac{\pi W^2}{8}$, The disk centre is above the surface substrate and at a distance $\sqrt{R^2 - \left(\frac{W}{2}\right)^2}$ from it. The radius R is obtained by solving the equation:

$$\frac{P_e Q_m}{\rho V} = \pi R^2 - \frac{1}{6W}(R(1 - \sqrt{1 - (\frac{W}{2R})^2})(3R^2(1 - \sqrt{1 - (\frac{W}{2R})^2})^2 + 4W^2)) \qquad \text{(V.4.2)}$$

This case can't be used to build a full density structure but can be interesting for a porous one.

Finally the clad geometry can be expressed as an analytical expression which is the one of the circle of radius R and centred at the distance $\sqrt{R^2 - \left(\frac{W}{2}\right)^2}$ below or above the substrate surface. The disk radius can be calculated knowing the width W and the powder efficiency P_e. It is to be noticed that the laser power is hidden behind P_e. We suggest two different methods to predict P_e and W.

Determining P_e and W with Parametric Relationships

Each clad is defined by the corresponding process parameters (P, Q_m, V). The previous parametric study can be used to link the clad cross section characteristics W, H and P_e to the process parameters. The geometrical form of the clad can now be predicted. This is presented in (Figure 29). The continue line represent the prediction from experimental measurements of the width and the powder efficiency. The dotted line is the prediction from parameters (P, Q_m, V) and the analytical relationships previously determined. It clearly shows that in the limits determined by our experiments we are able to predict the clad geometry from the analytical parametric relationships presented above.

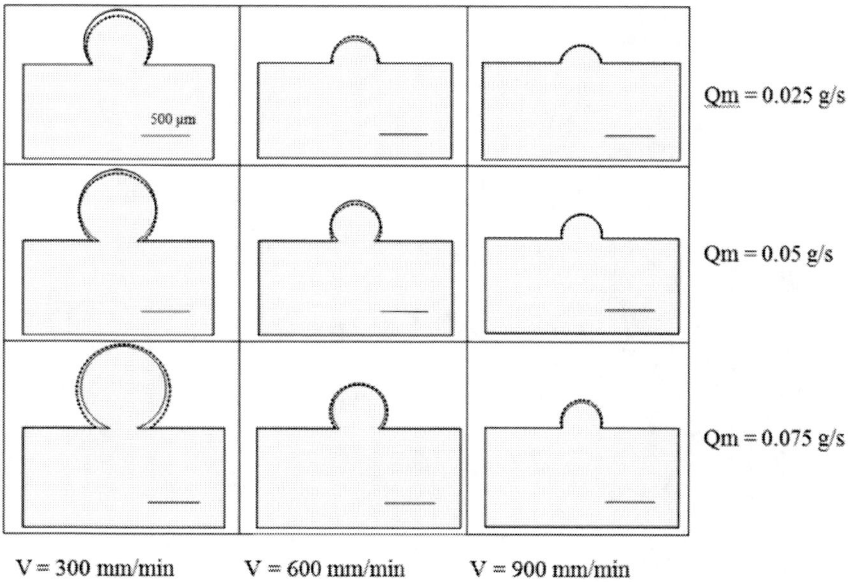

Figure 29. Confrontation between geometrical representations of the clad geometry obtained with experimental measures of W and P_e (continue line) and analytical determination of W and P_e from the parametric relationships (dotted line) at P=180 W.

In Figure 30 the cross section simulations obtained with the parametric relationships (dotted lines) are compared with those obtained with the experimental measured width (continue line) and superimposed with the experimental photographs.

These overlays are convincing, so this method can be used to predict the clad geometry and improve the process. The powder efficiency P_e is easy to measure weighing the part before and after the process. The width W can be easily measured as long as the clad doesn't correspond to the type 2 because of the narrowing bellow the clad upper part which imposes a cross section observation. A way to avoid this measure and to generalize the modelling is to determine the width W from a theoretically.

Determining W Using a Thermal Study

The relationship between W and the experimental parameters doesn't involve the parameter Q_m. This observation shows that a thermal process governs the parameter W. This is also in agreement with the usually accepted fact that only the powder which falls in the melting bath stay in the clad (Lin, 1999). The clad width is given by the molten pool dimension. The laser beam moves in the x direction at a constant velocity, many assumptions can be made to simplify the interaction between the laser and the substrate as follow:

- The substrate is considered as a semi-infinite solid initially at the room temperature T_0.
- The melting zone is supposed to be a sphere whose surface is always at the melting temperature T_m.
- Thermo-physical properties are temperature independent.

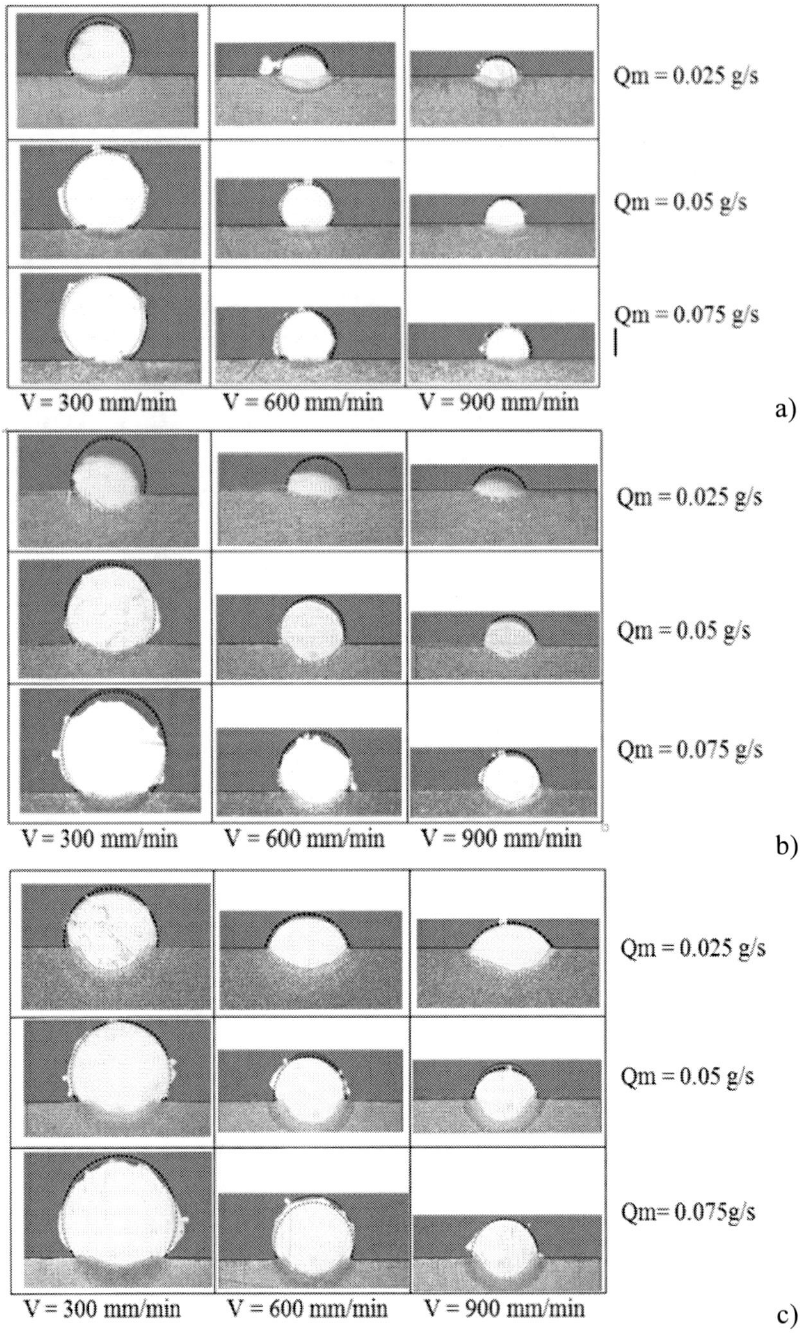

Figure 30. Confrontation between experimental results and simulated ones obtained with the analytical relationships at different power a-180, b-280 and c-360W.

In the steady state the pool geometry is constant. An energy balance is established between the absorbed laser energy in unit time, Q_l, and the energy transported through the boundary of the melting pool in contact with the substrate and the deposit clad in unit time Q_c by the heat conduction.

$$Q_l = Q \qquad \text{(V.4.3)}$$

The laser energy in unit time absorbed by the work piece is expressed as

$$Q_l = \text{Abs}\, P \qquad \text{(V.4.4)}$$

The proposed assumptions lead to a spherical symmetry problem solvable in one direction.

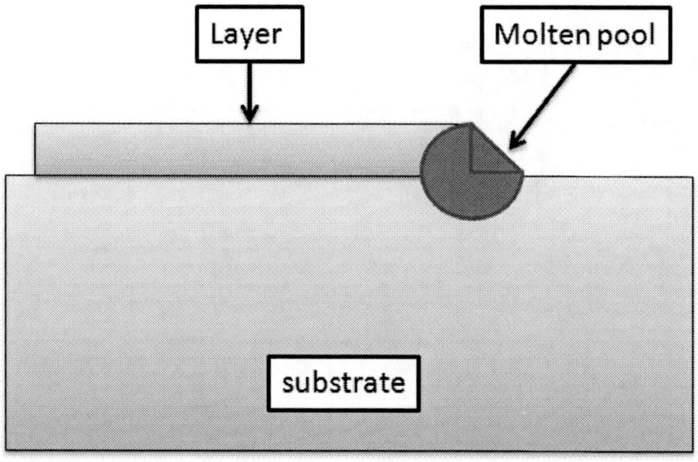

Figure 31. Schematic view of the spherical melting pool.

In these conditions and in case of transient conduction, an analytical solution is given by (Holman, 1990):

$$Q_c = \frac{k\,A\,(T_m - T_0)}{\sqrt{\pi\,\alpha\,t}} \qquad \text{(V.4.5)}$$

where A is the surface at the temperature T_m crossed by the heat flux i.e. the area of the melting pool in contact with the substrate and the deposit clad.

In laser cladding, the melting zone is almost spherical and the surface contact A represents a part of its surface as shown in (Figure 31), $A = f\,\pi\,d^2$.

The characteristic time for the melting zone to cross above a point is given as:

$$t = \frac{d}{v} \qquad \text{(V.4.6)}$$

From Equation (V.4.5) and Equation (V.4.6) the melting zone diameter is given by:

$$d = \left(\frac{\text{Abs}}{f}\right)^{\frac{2}{3}} \left(\frac{P\,\sqrt{\pi\,\alpha}}{\pi k \sqrt{V}(T_m - T_0)}\right)^{\frac{2}{3}} \qquad \text{(V.4.7)}$$

The molten zone width W is approximately equal to this diameter.

The width is given versus the melting temperature and the thermo physical proprieties of the clad material, and depends on the laser beam power and velocity.

However, powder efficiency depends on the particular nozzle used and seems difficult to generalize with an analytical relationship. The formula proposed by Liu and Li (2005) has been tried with our experimental results without success. So P_e is kept as an experimental entry. Results obtained with the analytical model are given in Figure 32.

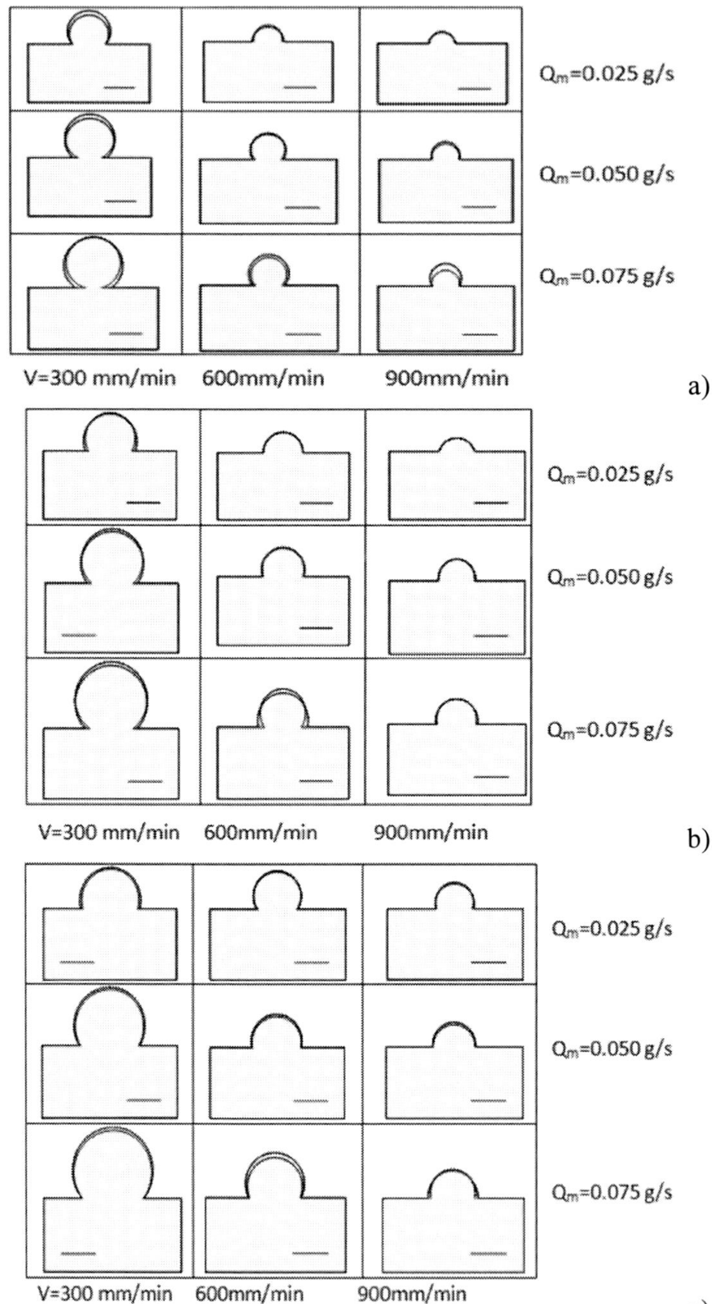

Figure 32. Confrontation between the representation of the clad geometry obtained with the analytical relationship (continue line) and with the thermal analytical model (dotted line) at different power, a-180W, b-280W and c-360W.

The continue line is the clad representation when W is calculated with the parametric relationships while the dotted line is the clad representation when the width W is calculated from the thermal analytical model. The visual confrontation shows a good agreement. So this model can predict the general geometry of the clad cross section. Changes induced by the parameters variations are well reproduced.

$$W = K.\left(\frac{P\sqrt{\pi\,\alpha}}{\pi k\sqrt{V}(T_m - T_0)}\right)^{\frac{2}{3}}$$

with K=Abs/f.

Absorption depends on the laser radiation, the used materials and the surface temperature. The fraction f can be estimated between 0.50 and 0.75. In our case, good results are obtained with a constant K equal to 0.26. So a calibration remains to be done.

Laser cladding process success depends on the ability to predict the laser clad geometry from the process parameters. Assuming the fact that the clad geometry is governed by the surface tension on the melted bath and so that the clad geometry is a part of a disk it is shown that it is possible to predict, with a good uncertainty, the complete geometrical characteristics of the laser clad. This modelling can also be useful in association with a more complete thermal study to determine the shape of the melting pool, and the temperature distribution in the building part.

VI. Thermal Study and Microstructure Prediction

VI.1. Thermal Study

Thermal simulations in laser processes are usually raised with the Finite Element Method very time consuming especially for a 3D problem. A simpler and faster alternative consists in using the point source method directly in the substrate surface. Regrettably the clad geometry is neglected and the calculated thermal field around the melting is too far from reality. In this part, a way to avoid this main defect is proposed. Temperature fields induced in the substrate by the laser source are calculated considering the laser beam as a point source applied on the building deposit surface. This is possible precisely using results presented above.

This approach requires several simplifying assumptions:

- The laser-powder interaction occurs at the clad surface extremity where it is being built. In reality, interaction between the laser and the powder has already started in the powder jet stream. An overestimated absorption coefficient compensates a part of it.
- The substrate is considered as semi-infinite body. This is reasonable consideringt the small size of the melting zone.
- The heat lost by convection and radiation is negligible compared to the supply of heat in the clad (Davis et al., 1986), (Cheng and Kar, 1997). Losses are generally considered underestimating the absorption coefficient.
- Thermo-physical properties of the part are considered independent of temperature.

- A steady state is established.
- The latent heat of fusion and vaporization are not taken into account, so temperatures will not be calculated in the melting zone.

The laser moves in the positive x direction with a constant velocity. The governing equation for the quasi-steady state heat conduction problem is given by

$$\nabla^2 T + \frac{V}{\alpha}\frac{\partial T}{\partial x} + \frac{q(x,y,z)}{k} = 0, \ -\infty < (x,y) < \infty \ \text{and} \ z < 0.$$

T is the temperature; q is the heat source, k the conductivity, α the thermal diffusivity and V the moving work piece velocity.

Boundary conditions are taken as

$$T(x,y,z) = T_0 \ \text{as} \ x \to \pm\infty; y \to \pm\infty, x \to -\infty .$$

T_0 is the ambient temperature.

The problem can be solved by using the Green's functions which satisfy the thermal equation:

$$\nabla^2 G + \frac{v}{\alpha}\frac{\partial G}{\partial x} + \frac{\delta(x-x_f)\delta(y-y_f)\delta(z-z_f)}{k} = 0, \ -\infty < x < +\infty, -\infty < y < +\infty,$$
$$-\infty < z < \infty$$

x_f, y_f and z_f are the position of the heat point source.

The Green's functions in the case of a quasi-stationary state are given by the following equation:

$$G(x,y,z) = \frac{1}{4\pi k}\frac{\exp\left\{\frac{V}{2\alpha}[(x-x_f) - \sqrt{(x-x_f)^2 + (y-y_f)^2 + (z-z_f)^2}\right\}}{\sqrt{(x-x_f)^2 + (y-y_f)^2 + (z-z_f)^2}}$$

For one point source heating the clad layer extremity, the temperature at a point (x,y,z) is given by:

$$\mathrm{T(x,y,z)} = \mathrm{T_0} + \frac{\mathrm{Abs\ P}}{2\pi \mathrm{k}}\frac{\exp\left[\frac{-\mathrm{V}}{2\alpha}((\mathrm{x}-\mathrm{x_f}) + \sqrt{(\mathrm{x}-\mathrm{x_f})^2 + (\mathrm{z}-\mathrm{y_f})^2 + (\mathrm{z}-\mathrm{z_f})^2}\right]}{\sqrt{(\mathrm{x}-\mathrm{x_f})^2 + (\mathrm{z}-\mathrm{y_f})^2 + (\mathrm{z}-\mathrm{z_f})^2}}$$

Only the part, in the substrate and the clad layer, has to be taken into consideration. A function, called z_f, depending on the space variables x and y is defined using the disk model to delimit the upper part of the substrate and the clad layer surface. This function $z_f(x,y)$ is so given by:

$$z_f(x,y) = \text{If}\left[\left(\sqrt{\left(x+\frac{W}{4}\right)^2 + y^2} < \frac{W}{2} \;\&\&\; x > \left(-\frac{W}{4}\right)\right), \sqrt{R^2 - \left(x+\frac{W}{4}\right)^2 - y^2 - \sqrt{R^2 - \left(\frac{W}{2}\right)^2}}, \text{If}\left[\left(\text{Abs}[y] < \frac{W}{2} \;\&\&\; x < \left(-\frac{W}{4}\right)\right), \sqrt{R^2 - y^2} - \sqrt{R^2 - \left(\frac{W}{2}\right)^2}, 0\right]\right]$$

where W is the laser track width and R is the circle radius. The source point is located at the clad extremity at a distance W/4 from the x axis centre.

The temperature distribution in the case of a laser power of 180 W, a powder mass flow of 0.025g/s and a velocity of 0.01m/s is shown in (Figure 33). The melting zone limit is defined as the isotherm 1535°C and the shown isotherms are $\{T_0, 100, 200, 300, ..., 1500, T_m\}$, where T_0 is the ambient temperature (T_0=20°C). Figure 34 shows the 3D melting zone geometry (Figure 34-b), and the volume in which temperature is higher than 600°C (Figure 34-a).

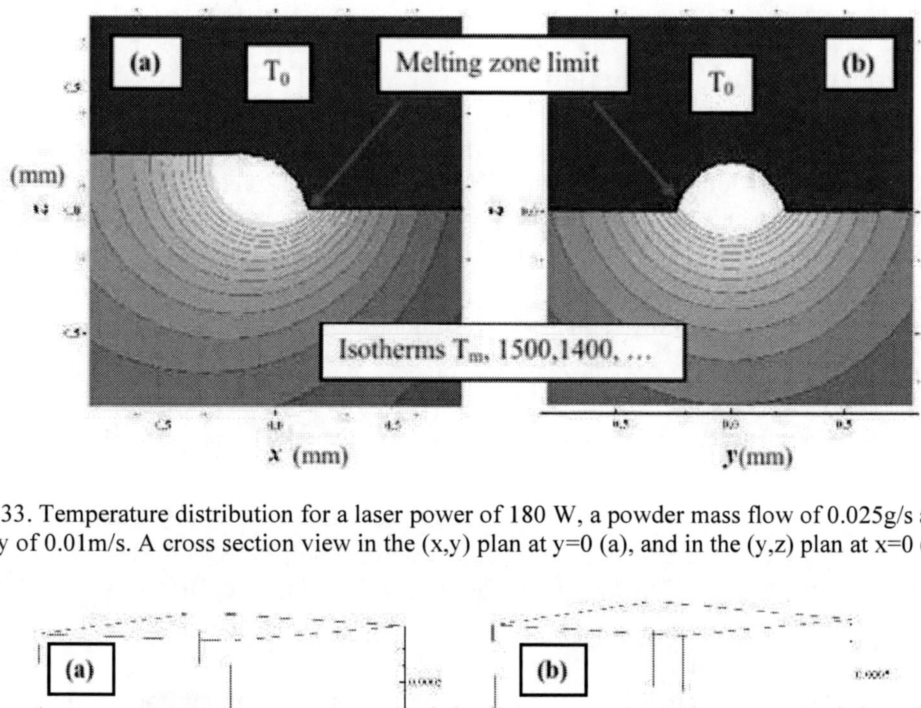

Figure 33. Temperature distribution for a laser power of 180 W, a powder mass flow of 0.025g/s and a velocity of 0.01m/s. A cross section view in the (x,y) plan at y=0 (a), and in the (y,z) plan at x=0 (b).

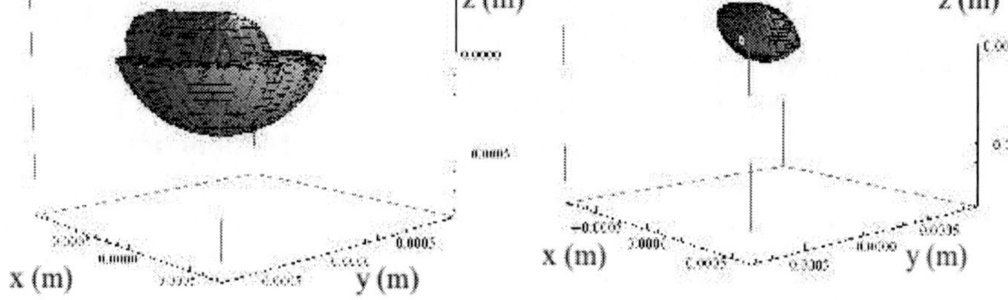

Figure 34. 3D temperature distribution for a laser power of 180 W, mass flow of 0.025g/s and a velocity of 0.01m/s. T>600°C (a), melting zone shape (b).

Absorption is fixed at 50% while the conductivity corresponds to the substrate conductivity 30 W/m/s. This choice is justify by the fact that heat is pumped from the melting zone by the substrate and so depends on the substrate conductivity. This modelling is not suitable in cases for which a narrowing is observed at the clad bottom because it should be as if heat should travel this empty space. Anyway, this configuration is not suitable for direct fabrication.

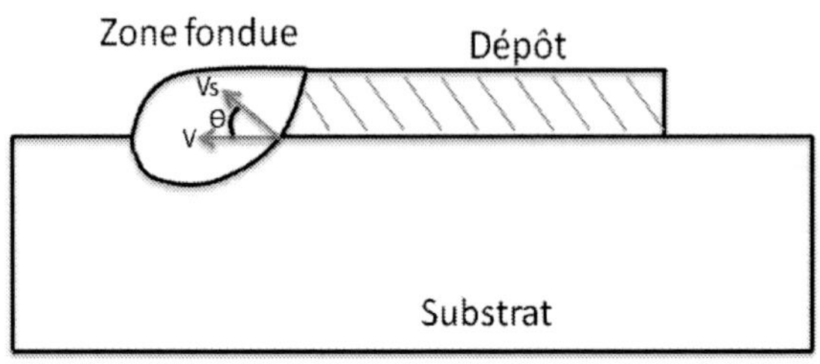

Figure 35. Solidification speed V_s and laser speed V at the solidification front level.

VI.2. Microstructure Prediction

Once the clad geometry and the thermal field in the building part calculated, the microstructure simulation seems possible. This part focuses on the confrontation between thermal simulations and the observed microstructures to understand variations of the grains morphologies and sizes. The microstructure characteristics depend on the solidification conditions in an extremely complex way. It depends on the physicochemical conditions prevailing in low space-time dimensions. Simulations require a precise knowledge of temperature gradients and solute concentrations in areas of transformations. However, a simplified and elegant approach is possible: the microstructure of the deposits is considered to be simply dependent on the temperature gradient ($g_T = \frac{\partial T}{\partial t}$) and the solidification front velocity V_S. These two quantities being considered in the normal direction to the solidification front (Kurz and Fisher, 1986). The quotient of the temperature gradient by the solidification front velocity governs the grains morphology, whereas the product of these two parameters governs the grains size. The solidification speed is related to the laser speed by the equation $V_S = V\, cos\theta$, where θ is the angle between the solidification front normal vector and the laser speed direction.

At the extreme bottom of the melting zone, the solidification front normal vector is perpendicular to the laser speed direction thus the speed V_S becomes zero. The ratio of the temperature gradient g_T on the solidification speed V_S (g_T/V_S) exceeds a critical value and then the grains growth is stable. The product ($g_T . V_S$) is very small which promotes the growth of large grains. A planar growth is then observed at the bottom of the molten zone as shown in Figure 36-a, b and Figure 37. At the top of the deposit laser clad, a part of the solidification front is perpendicular to the laser direction. The solidification front normal is then parallel to the laser moving speed and thus the solidification speed V_S is maximal and equal to the laser beam speed. The quotient (g_T/V_S) is minimal and the solidification front destabilization is maximum. This destabilization implies dendrites growth with potentially multiple branches. Elsewhere in the clad, the intermediate case must appear i.e. columnar dendrites growth. The columnar grains size depends on the process parameters and is always observed between 10 and 20 μm.

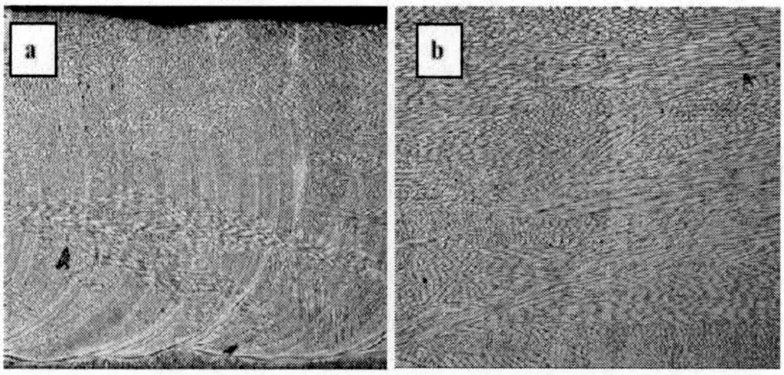

Figure 36. Longitudinal cross section for clad n 10 (general view x100 and detailed zone x200).

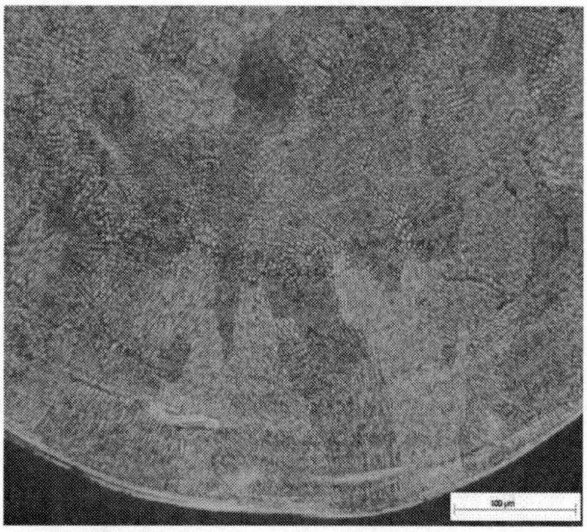

Figure 37. Cross section for clad n 16 (x200).

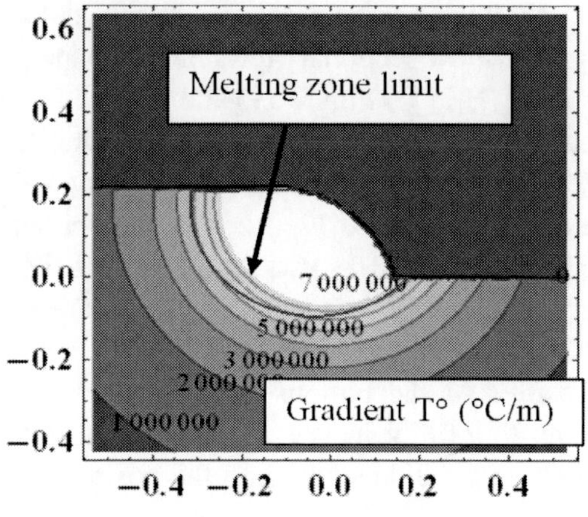

Figure 38. Temperature Gradients and melting zone limit for clad n 6. (P=180 W, Qm=0,050 g/s and V=0,01 m/s). Axis in (mm).

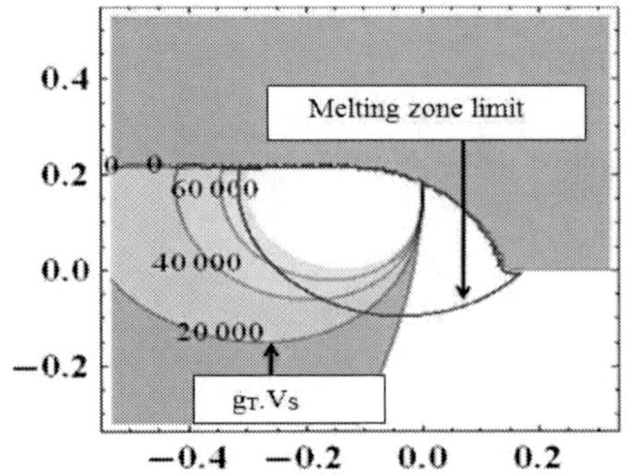

Figure 39. Product of the gradient temperature and the velocity of solidification ($g_T . V_s$) for clad n 6. Expressed in C/s.

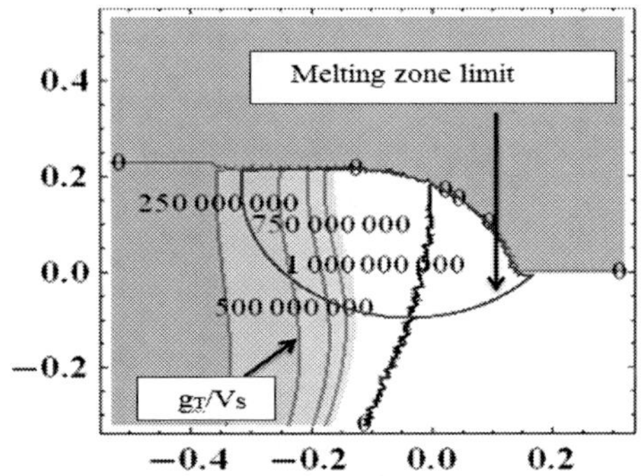

Figure 40. Quotient of the gradient temperature and the solidification front velocity ($g_T . V_s$) for clad n 6. Expressed in C.s/m².

Each process parameters influence is studied separately observing the deposits characteristics variations when one parameter varies while the two other ones are kept constant.

Simulations of the temperature gradients and the solidification speeds are caring out using the Green functions method in order to access to the product and quotient of these two quantities. The following figures show sectional views, in the plane y = 0, for the temperature gradient (Figure 38), the product ($g_T . V_S$) (Figure 39) and the quotient (g_T/V_S) (Figure 40).

Figure 38 shows the calculated temperature gradients around the melting zone. For the clad n 6, simulation indicates that the gradients are of the order of 5.10^6-7.10^6 C / m.

The product ($g_T . V_S$), exhibited in Figure 39, is zero at the bottom of the melting zone and gradually increases to reach over than 6104 °C/s at the top of the deposit. The grain size decreases when $g_T . V_s$ is growing and so the smallest grains are situated at the top of the molten zone. The quotient (g_T/V_S), exhibited in Figure 40, is minimal at the top of the molten zone with about 300.106 C.s/m². At the bottom of the melting zone, as the solidification velocity is zero, this ratio tends to infinity. This explains the well succession of different types of grains observed in the melting zone depending on the position of these grains in the clad deposit.

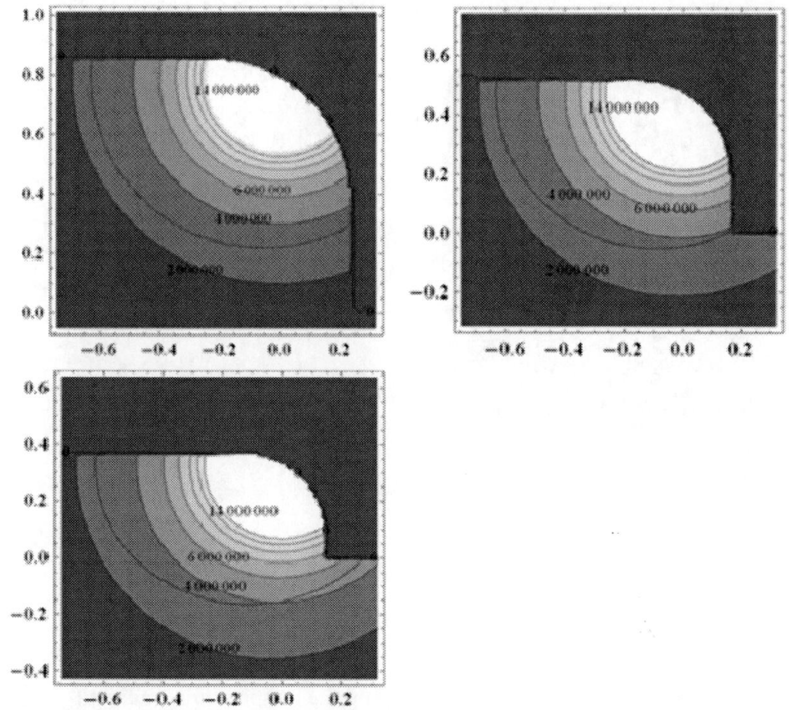

Figure 41. Temperature gradients and melting zone limit for clads n 13, 14 and 15.

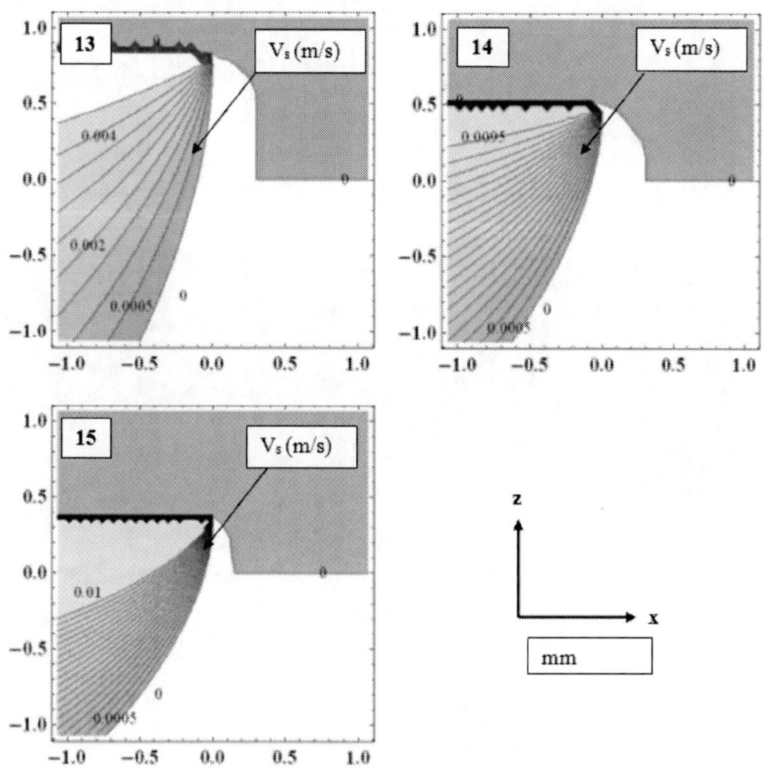

Figure 42. Solidification front velocity V_s for clads n 13, 14 and 15.

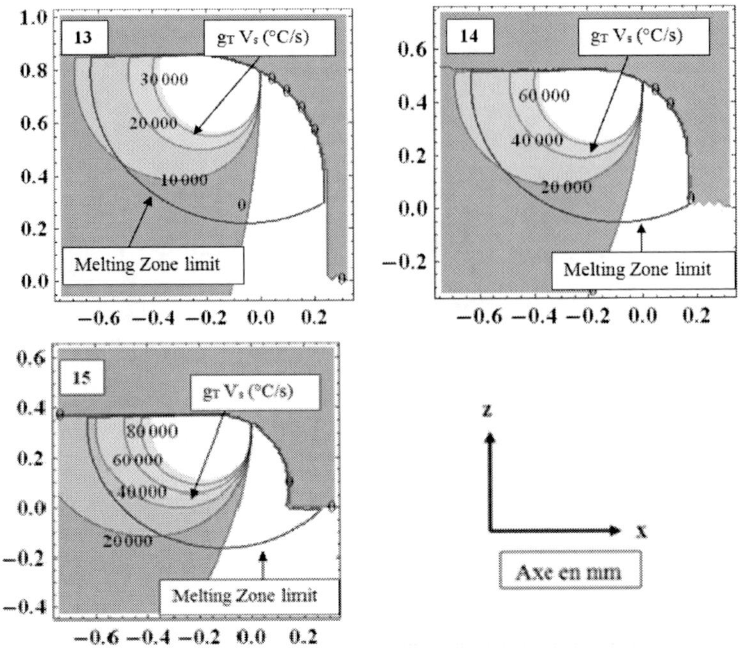

Figure 43. Product of the solidification front velocity and the temperature gradient ($g_T.V_s$) and the melting zone limit for clads n 13, 14 and 15.

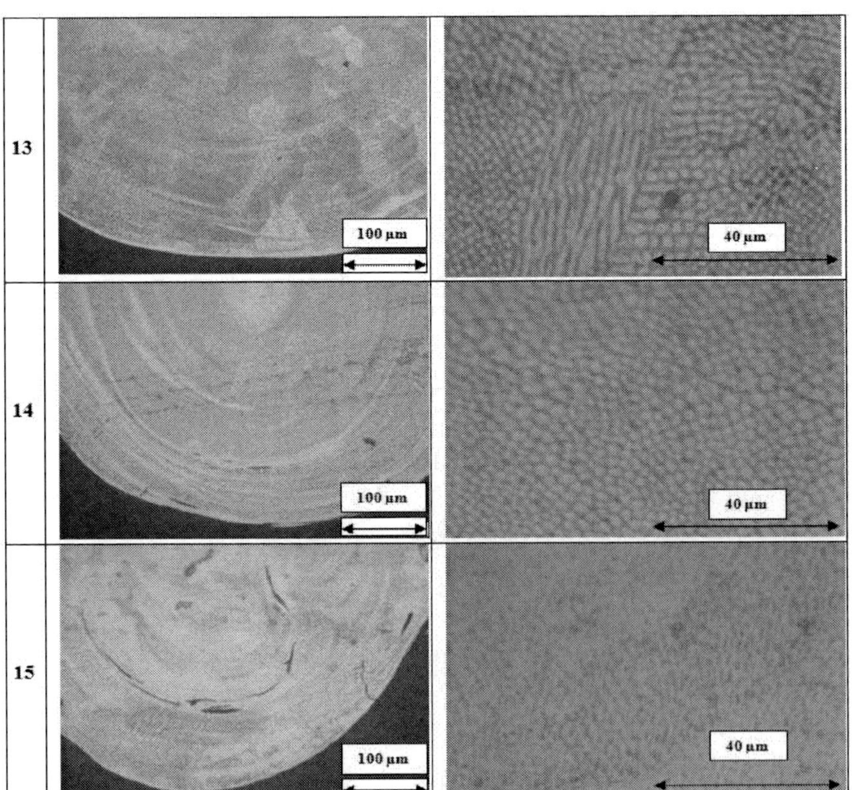

Figure 44. Observed microstructure for clads n 13, 14 and 15.

VI.2.1. Influence of the Laser Scanning Velocity V

When the laser scanning velocity increases, the solidification front velocity also increases as the melt follows the beam. Grains tend to become smaller as the speed increases as the product $g_T.V_s$ grows. This can be seen experimentally (Figure 44). For a constant power and mass flow, grains size is the tallest with a speed of 300 mm/min the smallest with a speed of 900 mm/min.

Figure 43 shows simulated results of the product of the solidification front velocity and the temperature gradient ($g_T.V_s$). Only the value of the product $g_T.V_s$ at the solidification front is interesting. This product is much greater at the top of the deposit, when the solidification speed V_s is equal to the laser speed. At this level, the calculated value is about 12000°C/s for the lowest speed, 25000°C/s for the intermediate speed and 38000°C/s for the fastest speed. When the scanning velocity increases the grain size should decrease. This result is consistent with (Kobryn et al., 2000).

Figure 44 shows three samples obtained with different velocities and the same power and mass flow. The three samples are the experimental witness of grain size variations with the scanning velocity and show that the grain size decreases when the laser speed increases. Note that with the strong variations in the value of this product in the same clad deposit we can observe different grain sizes in the same deposit (it varies from 0°C/s at the bottom of the molten zone up to several thousand °C/s at the top of the molten zone).

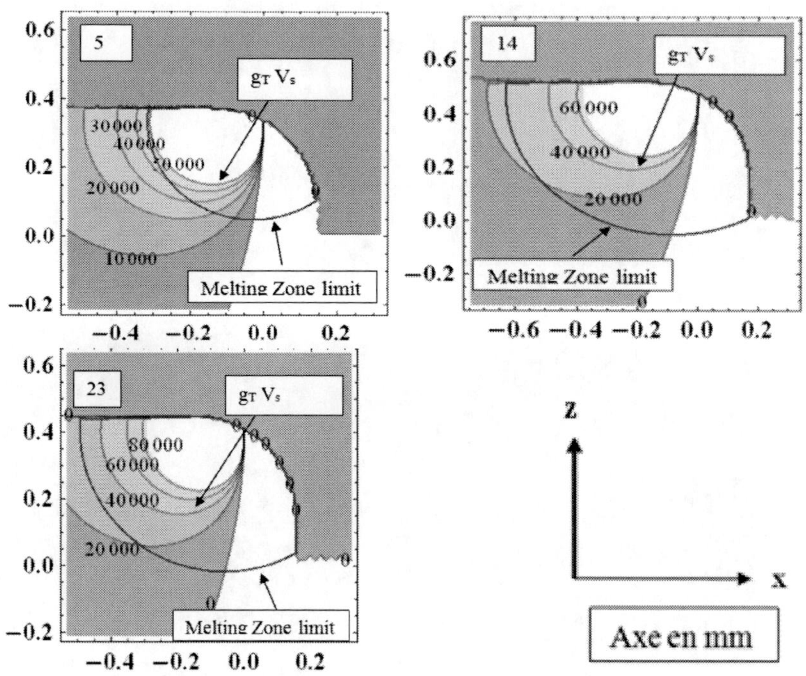

Figure 45. Product (g_T.Vs) and melting zone limit for clads n 5, 14 and 23.

VI.2.2. Influence of the Laser Beam Power P

Influence of the laser power P is studied on samples obtained with the same mass flow and velocity. The three deposits (5, 14 and 23) are built with a mass flow of 0.050 g/s, a speed of 0.010 m/s and the different powers 180, 360 and 280 W respectively. When the speed remains constant, the solidification front velocity is constant; consequently the product $(g_T.V_s)$ is governed by the temperature gradient. Our simulations show that increasing the laser power the molten zone becomes larger and the temperature gradients decrease near the solidification front. (Figure 45) shows the simulated product $(g_T.V_s)$. For a laser power of 180W (clad n 5), $g_T.V_s$ is around 50000°C/s at the top of the clad, for a power of 280W (clad n 23), $(g_T.V_s)$ is around 30000°C/s, and for a power of 360W (clad n 14), $(g_T.V_s)$ is around 25000°C/s.

For a constant speed and mass flow, grains in the clad obtained with a power of P= 180W are the smallest. Figure 46 shows the microstructure observed with an optical microscope.

VI.2.3. Influence of the Powder Mass Flow Q_m

Influence of the powder mass flow is studied on samples obtained with the same laser power and scanning velocity but different mass flows. Clads n 10, 13 and 16 have the same speed 0.005 m/s, same power 360 W and three mass flows equal to 0.025 g/s, 0.050 g/s and 0.075 g/s respectively. The scanning velocity remaining constant the solidification velocity is kept in the same interval. $(g_T.V_s)$ simulations (Figure 47) show that this product is constant on the melting zone boundary regardless of the powder mass flow sent. Grains size should not change between these three deposits. Figure 48 witness to the ineffectiveness of the powder mass flow on grains size.

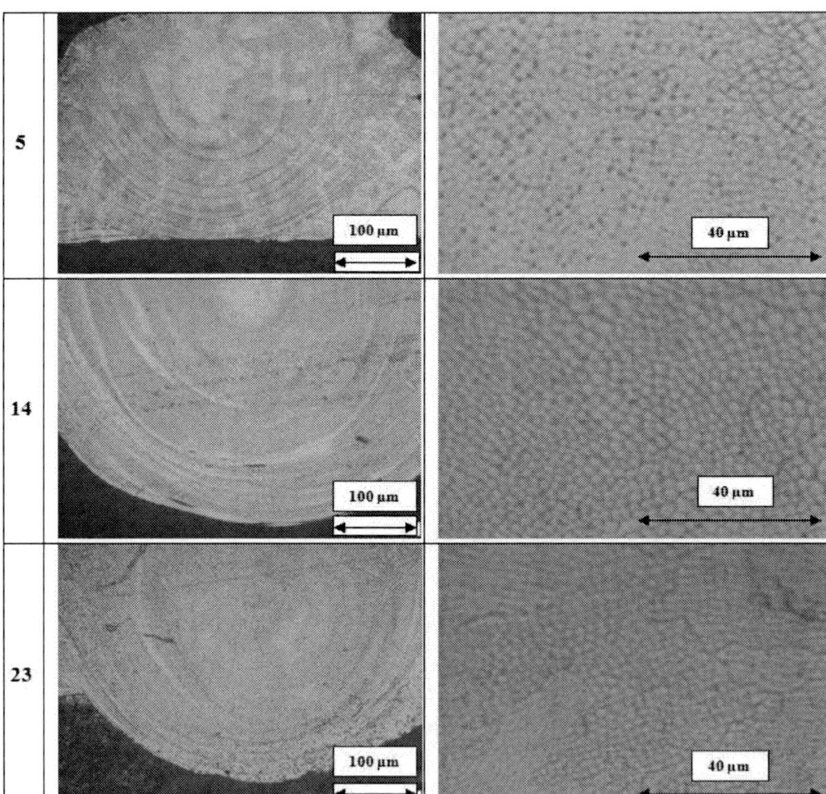

Figure 46. Observed microstructure for clads n 5, 14 and 23.

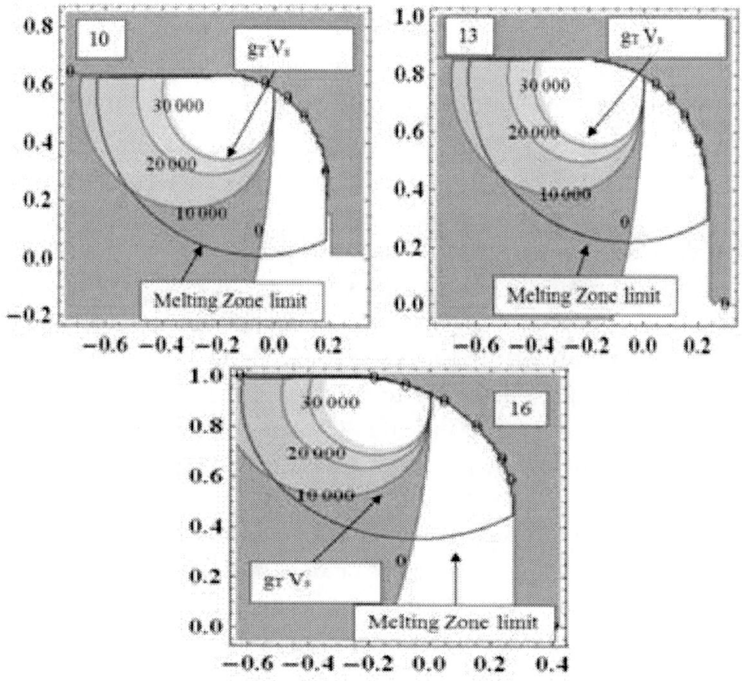

Figure 47. Product (g_T.Vs) and melting zone limit for clads n 10, 13 and 16.

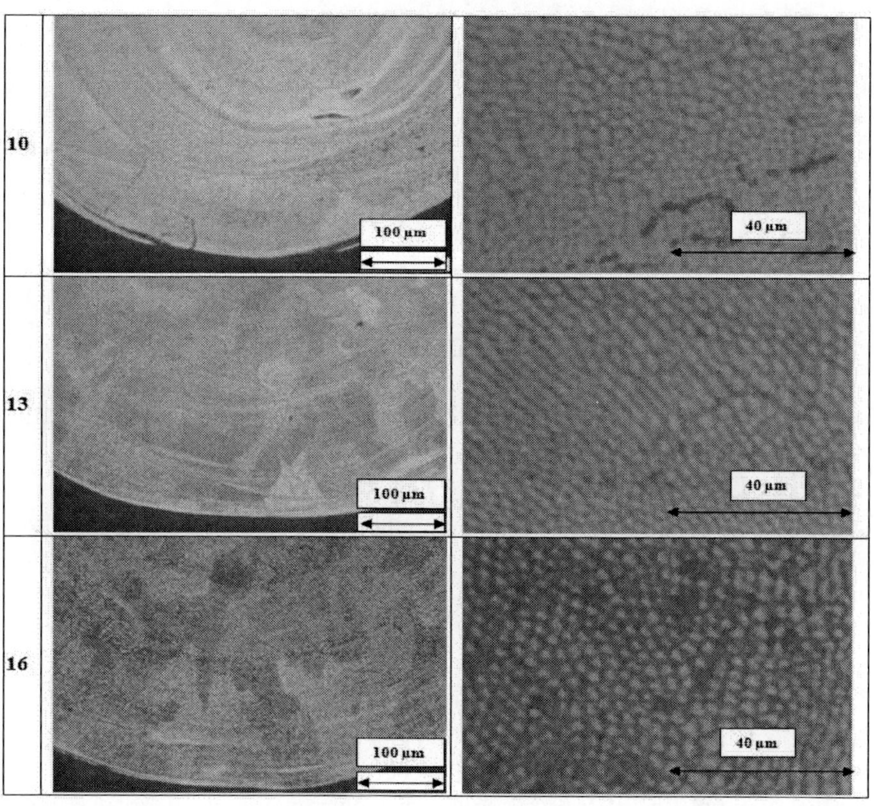

Figure 48. Observed microstructure for the clads n 5, 14 and 2.3.

CONCLUSION

In the present work, relationships such as ($P^{\alpha}Q_m{}^{\beta}V^{\gamma}$) between process parameters and cross sectional characteristics (height, width, area...) are established using the multiple regression analysis method. The good correlation coefficients obtained and the associated residuals analysis confirm the pertinence of such approach to predict the main geometrical characteristics of the clad geometry from the process parameters.

Prediction of the complete clad geometry is first investigated knowing the powder distribution in the jet. Three analytical functions are established to describe the clad geometry corresponding to three possible distributions. This leads to the conclusion that the powder distribution in the jet alone can't determinate the final clad geometry. Acting the fact that the laser clad cross section geometry is a disk due to surface tension forces; the centre position and the diameter of the disk are related to the clad width W and the powder efficiency P_e. Two ways are explored to calculate W from the process parameters. Finally the cross section simulations are compared to the experimental photographs. These overlays are convincing and so this pragmatic modelling is able to predict the complete clad geometry with an analytical expression. This is very useful to optimise the process parameters in order to build parts of good quality. The other interest point of this modelling is that it allows a very fast thermal study of the process in order to determine the melting pool shape and the temperature distribution in the building part.

A thermal study is proposed using the Green functions but taking account of the clad geometry. It can predict the melting zone limit and simulate the temperature gradients. The microstructure is also observed and compared with thermal predictions obtained simply from the product and the quotient of the gradient temperature g_T with the solidification front velocity V_S. Evolutions of the grains sizes and morphologies in the melted zone are explained by these calculations.

All these tools can be very useful to quickly take control of the laser cladding process or the Direct Laser Fabrication Process.

REFERENCES

Cheng, A. Kar. 1997. Mathematical model for laser densification of ceramic coating. *Journal of Material Science*, 32, pp. 6269-6278.

Colaco, R., Costa, L., Guerra, R., Villar, R., 1996. A simple correlation between the geometry of laser cladding tracks and the process parameters. *Laser processing: Surface treatment and film deposition. Ed Mazumder J, Conde O, Villar R, Steen W. Series E: Appl. sciences 307.*

Davis M., Kapadia P., Dowden J., Steen W. M., Courtney C. H. G. 1986. *J. Phys., D: Appl. Phys.*, 19, pp. 1981-1997.

De Oliveira U., Ocelık V., de Hosson J. Th. M. 2005. Analysis of coaxial laser cladding processing conditions, *Surf. Coat. Technol., 197, pp. 127-136.*

El Cheikh H., Courant B., Samuel B., Hascoët J. Y., Guillén R.. 2012. Analysis and prediction of single laser tracks geometrical characteristics in coaxial laser cladding process. *Optics and laser in engineering* 2012, 50, 3, 413-422.

Hoadley A. F. A., Rappaz M., Zimmermann M. 1991. Heat-flow simulation of laser remelting with experimenting validation. *Metallurgical Transactions*, 22B, pp. 101-109.

Hoadley A. F. A., Rappaz M. 1992. A thermal model of laser cladding by powder injection. *Metallurgical Transactions, 23B, pp.631-642.*

Holman J. P. 1990. *Heat transfer* 7TH *edition pp. 145.*

Kobryn P. A., Moore E. H. and Semiatin S. L. 2000. The effect of laser power and traverse speed on microstructure porosity and build height in laser deposited Ti-6Al-4V. *Scripta mater.*, 43, pp. 299-305.

Kurz, W. Fisher D. J. 1986. Fundamentals of Solidification *(Suisse: Trans Tech. Publications).*

Lin J. 1999. A simple model of powder catchment in coaxial laser cladding. *Optics & Laser Technology*, 31, pp. 233-238.

Liu J., Li L. 2005. Study on cross-section clad profile in coaxial single-pass cladding with a low-power laser, *Opt. Laser Technol.* 37 pp. 478–482.

Maiman T. H. 1960. Stimulated optical radiation in ruby. *Nature.*

Marsden C. F., Frenk A., Wagniere J. D. 1987. Power absorption during the laser cladding process. B. L. Mordike (Ed.), *Laser Treatment of Materials, DGM, Oberurel, pp. 375-380.*

Marsden C. F., Frenk A., Wagnière J.-D., and Dekumbis R. 1990. in Proc. *3rd Eur. Conf. Laser Surface, Treatment of Materials, ECLAT '90,* Bergmann H. W. and Kupfer R., eds., Erlangen, Germany, Sprechsaal Group, Coburg, Germany 1990, pp. 535-542.

Mazumder J., Kar A. 1987. Solid solubility in laser cladding. *Journal of Metals,* 39, pp. 18-23.

Mazumder J., Singh J. 1986. Laser surface alloying and cladding for corrosion and wear. NATO ASI Series, Series E: *Applied Sciences*, 115, pp. 297-307.

Onwubolu G. C., Davim J. P., Oliveira C., Cardoso A.. 2007. Prediction of clad angle in laser cladding by powder using response surface methodology and scatter search. *Optics & Laser Technology*, 39, 1130-1134.

Pinkerton A. J., Li L. 2004. Modelling the geometry of a moving laser melt pool and deposition track via energy and mass balances. *Journal of applied physics* D: applied physics, 37, pp. 1885-1895.

Toyserkani E., Khajepour A., Corbin S. 2005. Laser Cladding. *CRC Press LLC. Standard Book Number 0-8493-2172-7.*

Weerasinghe V. M., Steen W. M. 1983. Laser Processing of Materials. *Laser Cladding with Pneumatic Powder Delivery, pp. 166-174.*

In: Laser Technology, Applications and Future Prospects
Editors: D. Ebersbach and F. Gottschalk
ISBN: 978-1-62417-088-1

Chapter 4

LASER INDUCED TRANSFER OF ORGANIC MATERIALS

Dušan Kopecký* and Jitka Škodová
Department of Physics and Measurements, Institute of Chemical Technology, Prague

ABSTRACT

The development of next-gen micro-electro-mechanical systems (MEMS) like sensors, actuators, biocompatible implants, photovoltaic cells, batteries, or electronics circuits is often limited by process equipment, its accuracy, performance and applicability. Newly, advanced materials (fragile organics semiconductors, large macromolecules, polymers or even complex composites) are also used for these devices. These problems can create almost insoluble question of processing technology choice.

One of the possible solutions lies in the area of laser technologies adapted for transfer and processing of organic materials. Conducting polymers (polypyrrole, polyaniline, polythiophene), pthalocyanines, acetylacetonates, carbohydrates, Teflon, carbon nanoparticles and their composites etc. were successfully deposited on many types of substrates with defined thickness and shape without molecular decomposition using Pulsed Laser Deposition (PLD), Matrix Assisted Pulsed Laser Evaporation (MAPLE), Laser InducedForward Transfer (LIFT) or Matrix Assisted Pulsed Laser Evaporation – Direct Write (MAPLE-DW). Although these technologies are still far from use in mass production, they have key and irreplaceable role in experimental or even in manufacturing of prototypes.

This contribution brings general overview of laser technologies, in particular those used in transfer and processing of organic materials supported by author´s team experience and experimental results. This contribution is aimed especially at: principals of special apparatuses for PLD, MAPLE, MAPLE-DW using high energy lasers (working in UV, VIS and IR spectrum, both pulsed and continual mode), theoretical analysis of the molecular transfer mechanisms and optimization of the process parameters for non-destructive deposition. Furthermore, results of the molecular transfer based on electron microscopy and infrared spectroscopy will be discussed. Finally, some applications of mentioned laser technologies will be proposed.

* Email: kopeckyd@vscht.cz

INTRODUCTION

Lasers and Material Engineering

It has passed more than 50 years since construction of the first operating laser. During that time lasers expanded into many fields of human activities and today we meet with them almost every dayin electronics, optics, medicine, or in machining and measurement. From the beginning, however, lasers are closely linked to research. Material engineering can be considered as one of pioneer branches in the use of lasers. Material engineers have been evaluating results of interactions of highly energetic, monochromatic, coherent radiation with matter and have brought much new and fundamental knowledge about behaviors and properties of substances. In the very beginning, only the influence of laser radiation on the original and defined properties of irradiated materials was investigated – sometimes called primary effect of laser irradiation. Releasing of molecules by the laser irradiation from a target - so-called *ablation* (secondary effect of laser irradiation), which should be accompanied by condensation and subsequent formation of thin films (both effects together called as *deposition*), has become the subject of interest only later. A lot of scientific groups found that the deposited material is under certain - usually very specific – conditions, identical to the source material. On the other hand, in many cases, laser depositions lead to the creation of entirely new material structures and unknown metastable substances or sometimes also to total material destruction. Poor predictability of this behavior still attracts a large number of scientists. The application of laser deposition has been extended and modified in many different ways. Today it is already possible to nondestructively transport fragile molecules or molecular clusters of organic materials such as polymers, biocompatible materials and even living cells.

The main topic of this chapter is focused on technologies that were created by modifying the predecessor of all laser deposition technologies - pulsed laser deposition (PLD). The list of derived technologies would be too large and exceed this chapter. Therefore, this chapter deals only with laser deposition of organic materials. There are also other reasons for targeting this specific group of compounds. Organic materials are in many ways progressive substances with potential to bring new benefits or to completely replace some of the most widely used inorganic materials. Thus, technologies that will be suitable for the deposition of organic materials must be far more sophisticated to avoid damaging of more subtle molecules in comparison with inorganic counterparts. There is the power of laser deposition technologies.

Definition of Laser Deposition and Terminology

Technologies using lasers can be divided into *conventional laser processing* and *laser chemical processing* (Figure 1). The first group covers scribing, cutting, drilling, welding, annealing, recrystallization or glazing. It is usually used for physical treatment of material bulk. The second group includes surface doping, alloying, oxidation, nitridation, laser induced etching and finally *deposition.*This group is characterized by simultaneous changes in chemical composition of the material or activations of real chemical reactions [1]. Laser-based deposition technologies should be divided into methods with serial or parallel coating [2]. *Serial deposition* (Figure 2a) coats material sequentially; only a certain part of the substrate at a time is coasted; sometimes it is also called a patterning or laser printing. Typical example of this technology is Matrix Assisted Pulsed Laser Deposition – Direct Write (MAPLE-DW) or Laser Induced Forward Transfer (LIFT). On the other side *parallel deposition* (Figure 2b) coats the whole area of substrate with the material simultaneously. Patterning or laser printing using these technologies is also possible, but it is necessary to use special mask or shielding, which is usually impractical. Typical example of this technology is Pulsed Laser Deposition (PLD) or Matrix Assisted Pulsed Laser Evaporation (MAPLE). The most important types of laser technologies for transfer of organic materials are shown in the Figure 3.

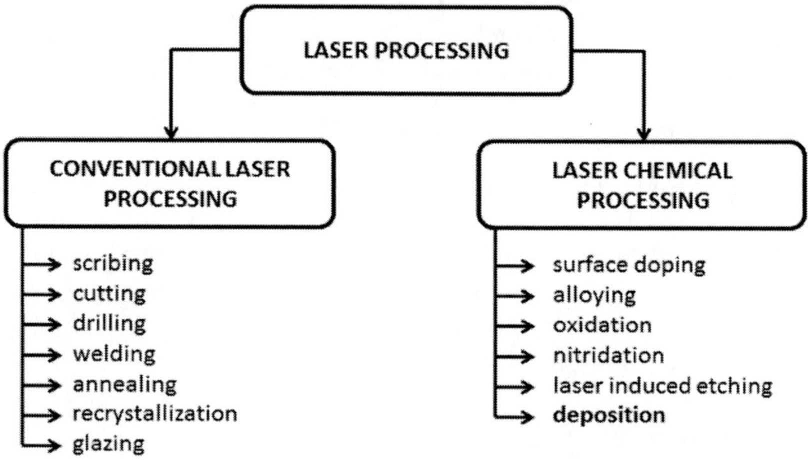

Figure 1. Types of laser processing.

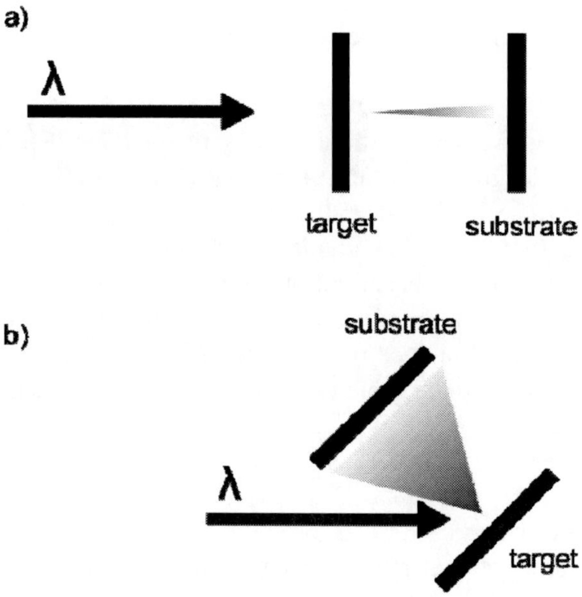

Figure 2. Serial (a) and parallel (b) laser deposition technologies.

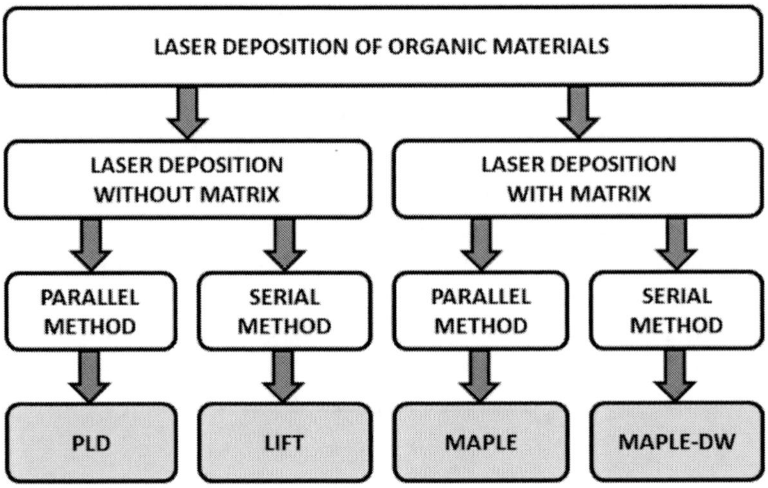

Figure 3. Types of laser deposition.

A lot of existing works about laser deposition use different terms or abbreviations, often describing the same technology. For clarification, this chapter is subdivided into three parts devoted to: (i) PLD, (ii) MAPLE, and finally (iii) both direct write technologies LIFT and MAPLE-DW. Other marginal technologies, terms and abbreviations will not be used, just some of them will be mentioned in very special cases.

Lasers used for Deposition

Laser is the main component, common for deposition apparatuses of all four technologies. Both parallel methods – PLD, MAPLE – use, without exception, only lasers working in pulsed mode. The main reason is to prevent overheating of the target and also to easy control thickness of prepared layer by using of defined number of pulses.

Serial methods – LIFT, MAPLE-DW - are less exacting and in general they use both laser modes – pulsed as well as continuous. Lasers used for deposition can be divided according to their wavelength. There are ultraviolet (UV), visible (VIS) and infrared (IR) lasers. It was found that the IR laser has different effects on the deposited material than the UV and VIS lasers. UV or VIS laser radiation produces more energy and the deposition is faster [3], but it also highly affects electronic states of organic molecules.

Given the fact that chemical bonds in most organic materials have energies well below the UV photon energies, this excitation may results either in direct bond dissociation (ablative photodecomposition) or in rapid conversion to vibrational energy with considerable overheating of the molecules (pyrolysis) [4,5]. On the other hand, IR laser mostly affects only vibrational modes of the organic materials. Thus ablation is gentler than that caused by UV laser [6].

General overview of lasers used in all four deposition technologies is shown in the Table 1. Further important laser parameters as fluence, pulse length, area of beam spot, number of pulses, beam repetition rate etc. will be discussed individually for every deposition technology.

Table 1. General overview of lasers commonly used in laser deposition technologies

Laser types	Wavelength	Spectrum	Depositions presented in this text
Solid-state laser			
Nd:YAG	266, 355, 534, 1064 nm	UV, VIS, IR	PLD, MAPLE, LIFT, MAPLE-DW
Nd:YAP	1078 nm	IR	LIFT
Nd:YVO4	266 nm	UV	LIFT
Er:YAG laser	2,94 μm	IR	MAPLE, PLD
Yb:YKW	1027 nm	IR	LIFT
Tm:Ho:LuLF	2 mm	IR	PLD
Molecular gas laser			
CO_2 laser	10,6 μm	IR	PLD
ArFexcimer	193 nm	UV	MAPLE, MAPLE-DW
KrFexcimer	248 nm	UV	PLD, MAPLE
XeClexcimer	308 nm	UV	PLD, LIFT
Free electron laser	2-10 mm		MAPLE

Deposition of Organic Materials

Deposition of organic materials is very difficult and complicated in comparison with deposition of inorganic materials. Inorganic materials are very robust, thus it is relatively easy to select a suitable deposition technique (magnetron sputtering, thermal evaporation, spin-coating, spray-coating and many others). Organic materials are often very fragile, have higher molecular weight and complexity of organic molecules, amount and type of side groups in molecule, different behavior in solvents and generally bad dispersing in solvents. All these characteristics influence the choice of the organic material deposition.

In the past, many technologies have been developed for deposition of organic materials, like organic molecular evaporation (OME), plasma polymerization, electro-chemical polymerization, spin-coating, electric-field-assisted CVD or photodecomposition [7]. Not all, however, have convincing results or desired precision.

Laser deposition technologies should be advantageous in case of: (i) deposition of thin films without affecting their chemical structure and physicochemical properties, with thin uniform solvent free coatings, (ii) necessity of high precision or sequential coating, (iii) deposition of multilayer structures [8], (iv) deposition of composites of organics/organics or organics/inorganics, (v) deposition of colloids or dispersion, (vi) deposition of structured materials, (vii) preparation of materials with better optical or electrical properties than initial properties of target material [9]. Indispensable advantage is also in localized effect of laser beam on target material. The mass of target is in comparison with other technologies like OME heated only locally for a short time in thin film near surface of a target, so the deposited material should be less influenced by overheating.

Organic materials discussed in this chapter are characteristic by: (i) affection to pyrolytic or photolytic decomposition, (ii) one or more properties of deposited materials are crucial for preservation of their function and a deposition should be concerned as successful only if these properties stay unchanged.

Application of Laser Deposition Technologies

Laser deposition technologies described in this chapter still belong rather to experimental technologies used for preparation of prototypes and material studies. However, a number of works from different areas of science and technology, which deal with these technologies, are outlining their future application possibilities. Such areas include micromechanics, metallurgy, integrated optics, electronics, semi-conductor industry, optoelectronics, sensor technology, chemical engineering, tissue engineering and production of implants, laser cleaning, biomedicine and others. Currently, there are the first prototypes of industry-usable apparatuses for material deposition by means of the PLD technology [10]. Some of these applications will be discussed within the following text, in connection with the particular technology.

Pulsed Laser Deposition (PLD)

Principle of PLD

History of Pulse Laser Deposition reaches back almost to the very beginnings of lasers. First experiments with laser deposition were carried out in 1965, i.e. mere 5 years after construction of the first usable laser. PLD is therefore belonging to the oldest and most popular technology of laser deposition of inorganic as well as organic materials. All other technologies of laser deposition originated as a gradual development and modification of PLD, with the aim to establish a technology more regardful of material sensitive to pyrolytic of photolythic dissociation.

PLD is based on a high-energy laser working in pulse mode. Laser pulses are hitting the surface of irradiated material – often called a target – from where they drive out a plasma cloud, which is then adhering to an opposite substrate (see Figure 4). Due to its arrangement, the PLD technology is classified as a parallel laser deposition method.

The mechanism of releasing the material from the target is based on transfer of laser beam energy to the ablated material and consequential explosive material driving-out from the target due to local overheating. At inorganic matters, partial pyrolytic and photolytic dissociation of material occurs in the point of local overheating, so that the emerging plasmatic cloud consists not only from molecules of deposited material and their clusters, but also from various fragments, electrons, etc. Subsequent condensing of the cloud on the substrate is accompanied by relevant chemical reactions and a film with molecules of the same stechiometric ration as that driven-out from the target is formed. It is obvious from the above mentioned, that this technology is suitable rather for deposition of inorganic materials. At deposition of fragile materials, thermal or photochemical damage to molecules occurs frequently.

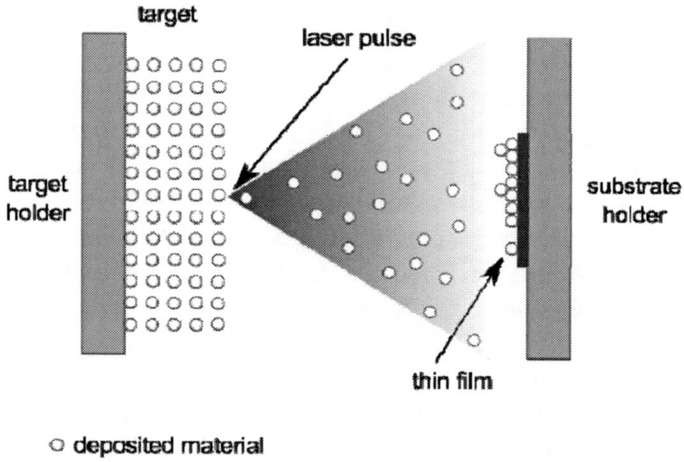

Figure 4. Principle of Pulsed Laser Deposition.

The Figure 5 shows an apparatus for deposition using the PLD technology. Laser beam is passing along an optical path consisting of a set of mirrors and lenses into a vacuum chamber (see Figure 5).

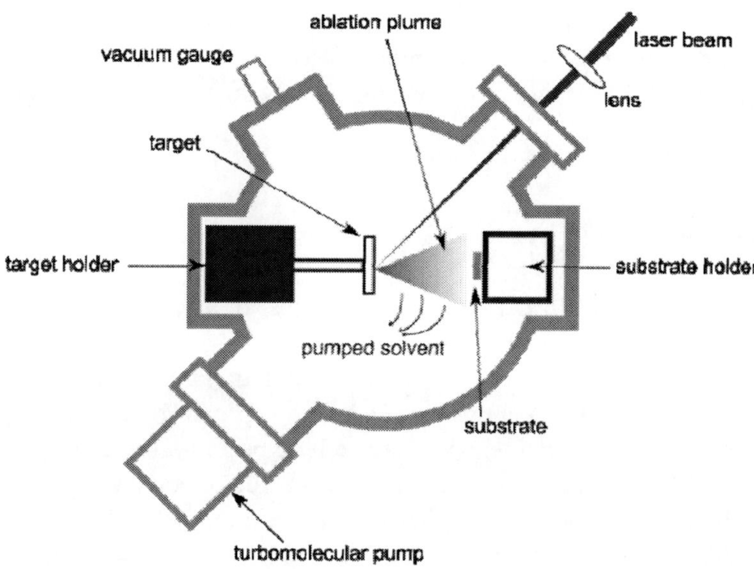

Figure 5. Scheme of PLD apparatus.

The vacuum chamber, which is generally evacuated by a rotary vane pump and a turbomolecular vacuum pump to high-vacuum pressure levels (HV), is increasing the medium free path of flight of ablated molecules of the target and at the same time it enables to create a residual working atmosphere of either inert character (inert gasses) or on the contrary it enables to carry out additional reactions (reactive atmospheres – oxygen, chlorine, etc.).

Within the course of PLD research, lasers from namely three areas of electromagnetic spectrum were tested – UV, VIS, IR. There is a principal difference in their effect onto material, in particular between the group of lasers from the UV and VIS spectra and those from the IR spectrum. At deposition with the PLD technology, laser irradiation affects the materials directly. The reason therefore consists in the mechanism of excitation of molecules of the irradiated target and thus in the mechanism of molecule ablation. While the UV and VIS lasers ablate quite quickly once the ablation threshold is overcome, however may excite electron states, the IR lasers may be a source of safer material deposition, since they cause only excitation of vibration molecule transitions. Moreover, due to some types of tuned lasers, we can target right absorption bands of specific couplings within a molecule. With regards to different effect, this technology is sometimes detached as an independent group and referred to as Resonant Infrared Pulsed Laser Deposition (RIR-PLD). In the consequence, the RIR-PLD deposition is nearing the laser induced thermal material evaporation, for example the deposition 5,5‘-di(9,9‘-di-(butyl)-fluorene‘)-2,2‘-bithiophene (DBFBT) with CO_2 laser [11].

The main advantage of the PLD technology is to deposit films with high repeatability and with accuracy within nm. In comparison to evaporation, where the composition of vapours of evaporated material depends mainly on pressure of vapours of the particular parts of the target, PLD is creating a cloud with stochiometric ratio of matters similar to that of the source target. This feature is principal mainly when depositing multicomponent materials and composites. Kinetic energy of deposited particles is such, that is enables their partial diffusion into the substrate surface. Therefore, deposited films excel in very good thermal adhesion. Further, setting of suitable deposition conditions (laser energy and substrate temperature in particular) enables to influence crystallography of the prepared films. In many cases it is possible to obtain material with new or improved optical or electrical parameters. A usual thickness of deposited film ranges over several place values from several nm up to several mm [7].

Chosen Parameters of PLD

A principal advantage of PLD is that a large number of parameters, which have influence onto the resulting parameters of the thin film, can be changed. The first parameter, already mentioned above, is laser wavelength, further length of laser pulse and/or its repeating frequency. At the same time, it is possible to control growth of the film by changing the target-substrate distance, substrate temperature or by type and pressure of gas in the residual atmosphere of the chamber [7]. The resulting films can be of epitaxy type or on the contrary can consist of large grain, those can be heterostructures, gradient films, they can consist of metastable materials, nanocystalic materials, composites, etc [1]. Some of the above mentioned parameters would be described in detail further. The fundamental parameters of deposition are mainly existence of so-called *ablation threshold* and *laser fluence*. The ablation threshold is such fluence, where the growth rate of the deposited film increases rapidly. The ablation threshold is dependent mainly on parameters and structure of the material, it is typical for given material and acquirement of it is fundamental for deposition. Figure 6 is showing simplified general dependence of growth rate on fluence. The turning point of this dependence is marking the ablation threshold. Generally, when using fluence below the ablation threshold, material deposition will occur, but the growth speed will be very low. On the other hand,7 high fluences above the ablation threshold will ensure fast deposition, but will also increase the probability of damaging the structure of the transferred material. As a conclusion it is obvious, that optimal energy of laser radiation is right within the vicinity of the ablation threshold. Some works even show that in case of some materials there are just very narrow intervals of fluences, which enable depositing the given material. I.e. Lösekrug et al. [12] states, that for polymethylmehycrylate (PMMA) deposition using the KrF excimer laser, only a narrow range of fluences ranging from 80 to 160 mJ per square cm can be used; this range is found quite close to the ablation threshold (60 mJ per square cm).

Laser radiation pulse length – this is a parameter, which may substantially affect mechanism of deposition. Well-chosen length of laser pulse is influencing the temperature effect acting onto the target of source material and thus it is possible to prevent damage of the chemical structure of the deposited matter. An example can be using of femtosecond laser. Hopp et al. [13] claims, that using of femtosecond laser at deposition of organic matters may substantially increase the share of transferred matter without damage to its structure.

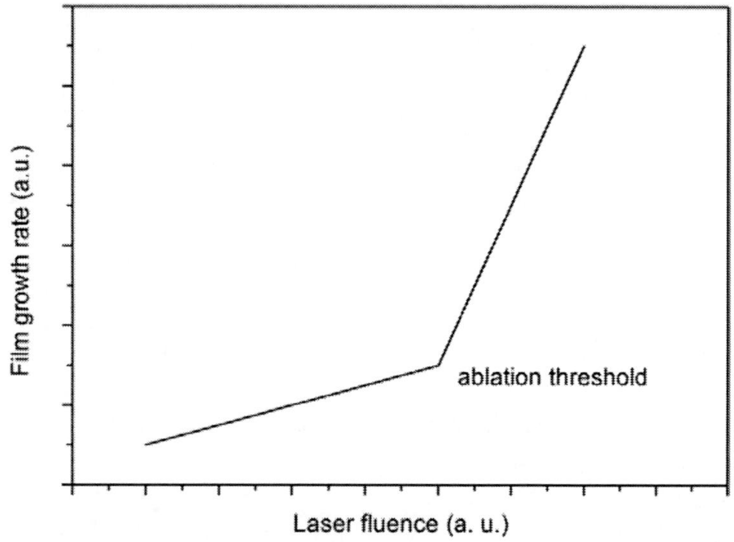

Figure 6. Simplified dependence of film growth rate on laser fluence for a general material.

Hopp has suggested a simple deposition mechanism for femtosecond or picosecond lasers, involving dissociation of thin surface layer of material. The material being explosively released from the layer, tears off further molecules or even whole grains of material by means of recoil forces, and these are transferred without any contact with the laser radiation (see Figure 7). This work deals with using lasers for deposition with ultra-short pulse time, which causes lowering of ablations threshold up to two place values. As a result, only a very thin surface layer of the target is influenced and there is only minimal damage to the neighbouring material.

Another, in literature little mentioned parameter, is *mutual arrangement of the substrate and the target*. Wisz et al. [14] received interesting results at deposition of pentacene onto a substrate placed horizontally or vertically towards the target (see Figure 8).

According to his conclusions, molecule diffusion into the substrate surface occurs on the horizontally placed substrate, and nucleation centers are formed after exceeding certain critical concentration. Other pentacene molecules are consequently caught by these nucleation centers and structured crystal growth takes place.

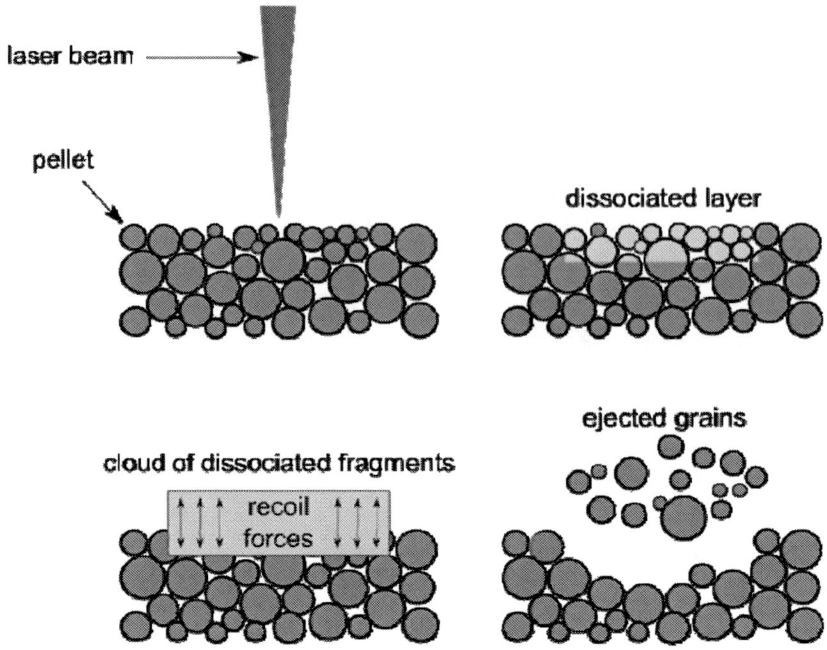

Figure 7. Principle of PLD using femtosecond laser, according to [13].

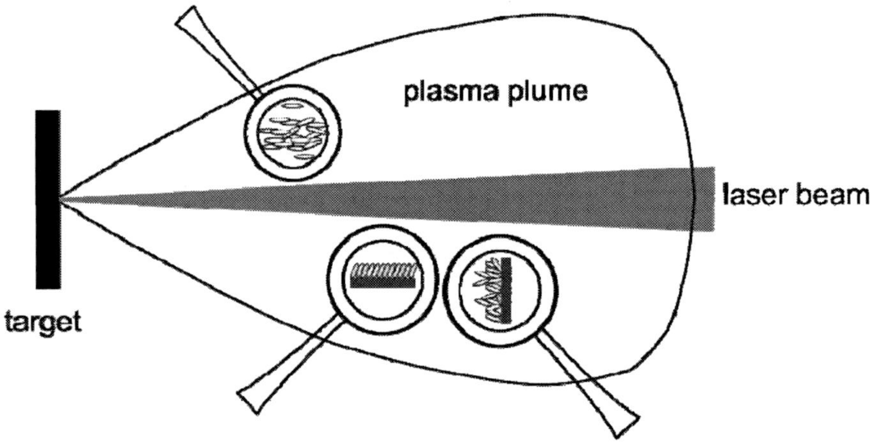

Figure 8. Film growth on horizontally or vertically placed substrate, according to [14].

On the other hand, film growth on the vertically placed substrate is more complicated and it is characterized by crystal growth on current nucleation centers and simultaneous formation of new ones, which causes a dendritic structure to be formed.

Method of target preparation, its temperature during the deposition and *substrate temperature* belong to other important parameters. Classical method of target preparation for PLD consists in pressing of powder of source material in a pressing matrix. Powder pressing reminds of metal sintering, in certain cases even increased temperature is used to ensure partial smelting of the source material and formation of smooth surface without air pokes [15]. A target can be made from blocks of commercially supplied clean materials as well.

A target placed into a vacuum chamber is usually left at the ambient temperature. In special cases, the target can be further heated up, mainly in order to reach faster film growth. Such increase will be evident especially in cases, when laser energy is far under the material ablation threshold. Under certain conditions it is advantageous to heat up also the substrate. Increased temperature is here used namely for modification of crystallinity of deposited film, which occurs in particular when we are moving around glass transition temperature or melting-point temperature [16].

Deposition of Organic Materials by PLD Technology

The main attribute of PLD is direct effect on high-energy laser radiation onto target material. In this respect, using of PLD at deposition of organic matters is very limiting, since there occurs at least partial dissociation of chemical structure of the source material. Experiments carried out in the course of several tens of years have however shown that PLD may be in certain cases fully suitable even for deposition of organic materials.

In general, we can discuss three categories of matters for which PLD is suitable: (i) the first category consists of matters with ability of self-composition after decomposition by laser radiation, a classical example of this group is polytetraflouroethylene (PTFE); (ii) to the second category belong all matters – polymers, macromolecules and biologically active matters – where laser transfer causes partial decomposition, but the quantity and extent of the damage does not affect its function; (iii) in the last case the material degradation during deposition is apparent and it concerns majority of transferred molecules, but it is of no inconvenience as far as the goal is preparation of metastable material, segments of molecules with new parameters or even the study of pyrolytic decomposition of matters [17].

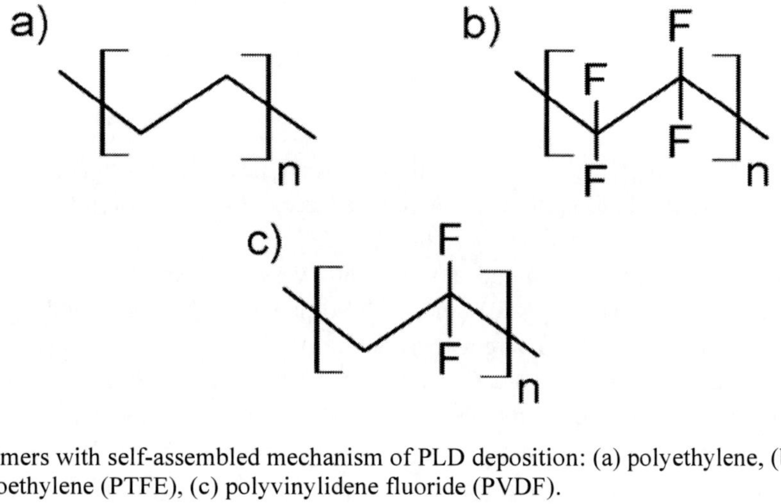

Figure 9. Polymers with self-assembled mechanism of PLD deposition: (a) polyethylene, (b) polytetrafluoroethylene (PTFE), (c) polyvinylidene fluoride (PVDF).

As above-mentioned already, deposition of PTFE and several other simple, linear polymers belongs to the first category. PTFE (Figure 9b), also known under its trademark Teflon, is a fluorized polymer with excellent mechanical, thermal and chemical stability, low dielectric constant, surface adhesion and low friction resistance. Due to its features it is used in vast number of applications, such as in microelectronics, mechanical engineering, medicine or chemistry and therefore mastering of preparation of PTFE thin films is crucial. Standard deposition technologies however face the problem of its poor solubility in solvents, therefore the PLD technologies seems to be a suitable solution for preparation of thin films [18].

One of the first research studies of PTFE deposition with the PLD technology was carried out by Blanchet et al. [4,19,20]. He proposed the mechanism of PTFE deposition as its pyrolytic decomposition due to fast change of excitation of electronic states onto excitation of vibration states, followed by local overheating and depolymerization. Ablation cloud of formed monomers and fragments then repolymerizes after incidence with the substrate and it forms a thin film.

This unique mechanism was observed at PTFE [15] and several structurally very similar polymers, such as polyethylene (PE) [21], polyvinylidene fluoride (PVDF) [16] (Figure 9), however it was never observed at any other group of matters, i.e. at macromolecules or biologically functional complexes.

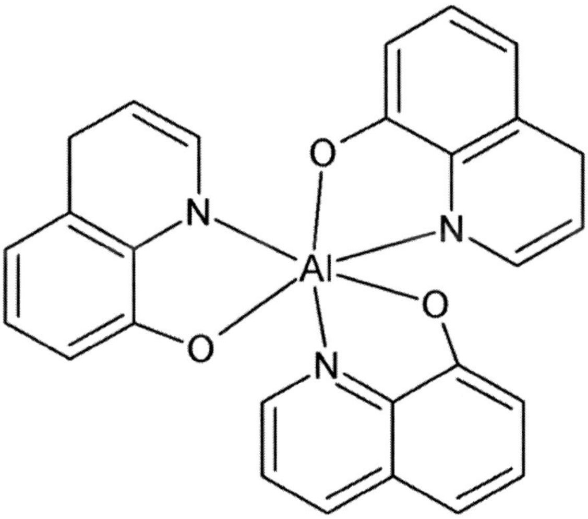

Figure 10. A representative group of macromolecules - tris-8-hydroxyquinline (Alq3).

Matters, whose integrity may be partially disturbed, nevertheless the matter does not lose its functional features, belong to the second category. For an example may serve some electroluminescent or biologically active matters.

The following matter – to name a few – are very well examined: aluminium tris-8-hydroxyquinline (Alq3) (Figure 10) and N,N'-diphenyl-N,N'-bis(3-methylphen-yl)-(1,1'-biphenyl)-4,4'-diamine (TPD). There is certain limitation at deposition using the Nd:YAG laser (355 nm), consisting in the necessity to use preferably low fluence (under 25 mJ per square cm) [22] or not to exceed the ablation threshold which lies roughly in tens of mJ per square cm; similar results are reported in [23,24] as well. At such energies, the main mechanism of transfer is slow desorption rather then ablation [25].

Another telling example of a matter belonging to this category is pepsin. This quite sizable and complicated molecule with a number of side groups was deposited by two independent research groups. Successful transfer was verified by an ability of a thin film of pepsin to fissure the protein ovalbumin [26], which was confirmed in both cases though certain marks of light modification of the enzyme structure [27,28]. Similar results were reached using a femtosecond laser [13]. At some biologically active matters, however, even a slight change in structure may cause disruption of its function. The example is deposition of urease with the PLD technology, which causes its damage. The MAPLE technologies proved to be a more suitable solution.

Another of representatives of biologically active matters, biopolymer polyhydroxybutyrate deposited with the ArF excimer laser with wavelength of 193 nm shows at low densities of fluence (0,12 J per sq.cm) correspondence of FTIR spectra with the original source material of the target. Higher fluences manifest themselves negatively [7]. Polyhydroxybutyrate can be classed into the category of matters showing after PLD deposition partial damage without principal influence onto functionality. This category includes also albumin [27], polyepichlorhydrine [28], collagene and keratin [29], poly(p-phenylene sulfide) (PPS) [30,31], poly(3-octylthiophene) [32], pentacene, thienylene–vinylem [33], and others, which show signs of partial damage with no significant influence onto functionality. We cannot however ignore, that some of the works state that the best deposition conditions are found at low densities of laser radiation energy, where the deposition is controlled by slow desorption rather than by ablation.

The third category includes all other matters, which yield to pyrolytic or photolytic decomposition and where irreversible loss of function takes place. In certain cases, however, pyrolytic decomposition may be the goal. One such example is study of laser stimulated pyrolytic decomposition of the poly(arylene ether-sulfones) (PES) group, in order to compare the mechanism of the molecule disintegration with those at classical pyrolysis [17].

Another example may be deposition of indium acetyl-acetonate (InAcAc). The InAcAc molecules indeed do disintegrate during the deposition, however the resulting film is getting new features suitable for example for ozone detection [34].

Similar behaviour is found at poly(ethylene glycol) (PEG) [6] or polymethylmethacrylate (PMMA). Both are getting modified at the PLD technology, target fluxion occurs and suitable fluence to make possible material ablation [12,35] has to be sought for. In spite of these limitations, PMMA can be further used in composites as a biocompatible matter.

Matrix Assisted Pulsed Laser Evaporation (MAPLE)

Matrix Assisted Pulsed Laser Evaporation (MAPLE) technology, developed in Naval Research Laboratory [36], features substantially finer ablation mechanism in comparison with PLD. It enables forming of thin and uniform films of material easily liable to photolytic or pyrolytic decomposition. MAPLE is therefore used for deposition of organic and biological materials with character of polymers, macromolecules or biologically functional complexes (including living cells).

The finer mechanism of ablation is due to different constitution of the irradiated target. The target consists not only of deposited material, but also of so-called matrix. Matrix is a matter of low molecular weight, whose absorption band has ideally its maximum at the wavelength of used laser radiation. Laser radiation hitting the target is then absorbed mainly by the matrix, which acts as an effective shield protecting the deposited material from photochemical damages. As matrix, volatile organic solvents of the deposited matter are most frequently used (such as methanol, dimethylsulfoxide, toluene, etc.), but also water. Most of organic matters show also high absorption bands in the UV area. For this reason, typical concentration of deposited material in matrix is approximately several % of weight [2,5], which minimizes the probability of interaction of deposited material with UV radiation. When choosing the solvent, possible photochemical decomposition of the solvent must be taken into consideration. Selection of improper matrix does not have only to cause damage to the material due to photochemical or pyrolytic degradation. For example, Caricato et al. [37] states that in the case of deposition of poly(9,9-dioctylfluorene) from chloroform at 248 nm, photochemical disintegration of chloroform and formations of radicals may occur, which consequently causes damage to the deposited material; tetrahydrofurane and toluene on the contrary do not show such behaviour. Tunny et al. [37] has the similar experience with chloroform.

Target for irradiation is made by freezing a solution or dispersion of deposited matter in matrix to solid state, at the temperature of liquid nitrogen. Deposition itself takes place in evacuated chamber, similar to that of PLD, see Figure 5. Laser beam is hitting the target under angle of 45°; the arrangement corresponds to that of parallel deposition method (Figure 2b). After the laser beam hits the target, energy of laser pulse is absorbed by the matrix. This phenomenon can be described as photothermal process. Energy of photons, which is absorbed by the matrix, is changing to thermal energy and that is further causing intense local overheating of the matrix and its evaporation to gas phase. Deposited material, protected from laser radiation by high absorption ability of matrix, is receiving sufficient kinetic energy by collective collisions with its molecules, which enables formation of cloud of ablated material and its transfer onto substrate [2,5] and homogeneous film of deposited material is being formed on the substrate surface. Matrix is quickly evaporating due to low vapour pressure and it is continuously pumped-off by turbomolecular vacuum pump, so that there is no rise of pressure within the chamber. Under certain conditions it is suitable to carry out deposition in so-called residual working atmosphere, which may consist of i.e. inert gas He, Ar, N_2 preventing oxidation of deposited material, or on the contrary in reactive atmosphere, which enables chemical modification of deposited material during deposition.

Since the deposited material is highly diluted in the matrix, MAPLE technology shows substantially lower consumption than PLD, but the deposition process is slower [39]. In an ideal case the target consists of true solution; due to generally worse solubility of many polymers and macromolecules it is possible to work also with colloidal solutions or dispersions, however solid particles must be perfectly scattered within the matrix. Parameters of the matrix significantly influence resulting chemical composition and morphology of the thin film. When choosing the matrix, surface wettability of the desired solute material, solubility, absorption coefficient of the solvent material at the laser wavelength, solution vapour pressure, laser penetration depth, solution freezing temperature, and heat capacity [40] must be taken into consideration.

Pulse lasers with wavelength in the UV, VIS or IR range of spectrum are used for depositions. Choice of wavelength depends on type of deposited matter and namely on used matrix, since solubility of many matters may be restricted by limited number of solvents. There is similar rule for choosing the wavelength as for the PLD technology: with increasing wavelength, probability of material photochemical decomposition diminishes, but growth rate of the film is negatively affected. In certain cases, using lasers with tunable wavelength [41] seems to be advantageous. Survey of used lasers is shown in Table 1.

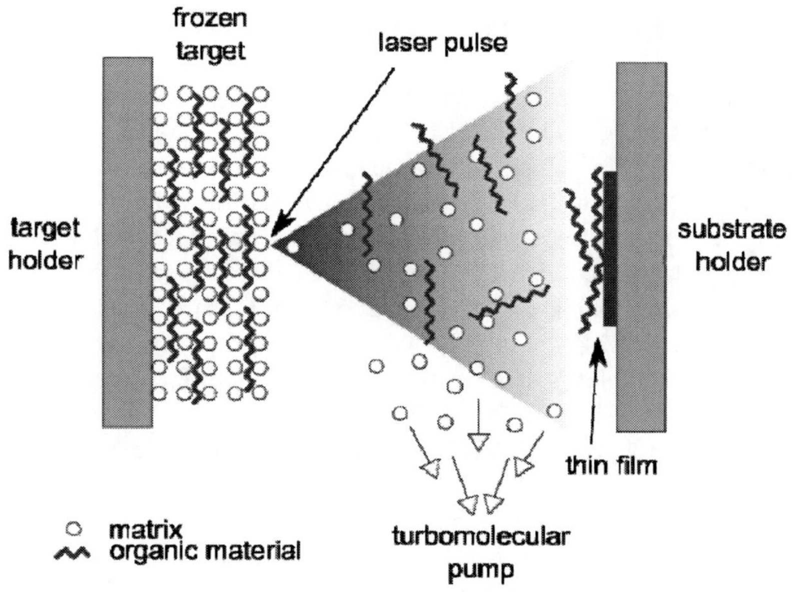

Figure 11. Principle of Matrix Assisted Pulsed Laser Evaporation.

Pressure within the working chamber is kept around 10^{-3} Pa. Pressure fluctuations during deposition can influence morphology of resulting thin films, therefore a powerful set of vacuum pumps is connected to the chamber to keep low pressure values not only before the deposition, but also during its progress.

In the vacuum chamber, a target holder is placed, which enables cooling of the target during the whole deposition process. The target is often rotating during deposition to provide even and homogenous ablation of deposited materials from the whole target surface. So-called crater effect occurs at stationary targets, which decreases deposition rate. Opposite to the target, a substrate holder is located.

For experimental purposes it is usually made of polished silicium, plastics or glass. New works from medicinal area are using various implants or medicaments as a substrate. Measuring of thickness of emerging film takes place either in situ by means of micro-scales based on quartz crystal microbalance (QCM) or ex situ using a Talystep. Measuring in situ enables precise regulation of films thickness by controlling the number of laser radiation pulses. The MAPLE technology enables to prepare films of thickness ranging from 10 nm up to several μm; in case of using different targets also multilayers consisting of various materials [8].

Quality of thin film prepared by the MAPLE technology is influenced by numerous parameters. Among the most important ones, there is already mentioned laser wavelength, number of laser pulses, type of matrix, concentration of deposited matter within the matrix, target-substrate distance and working pressure [42]. Another parameter, which influences the thin film structure, is fluence. This parameter has however smaller influence onto the deposited material structure than with the PLD technology, providing a suitable matrix is chosen. Substrate temperature belongs to another parameters influencing morphology of the emerging film [40,43,44].

Deposition of Organic Materials Using the MAPLE Technology

Efficiency of matrix at laser deposition is evident when comparing results of deposition of the same molecules deposited by the PLD and MAPLE technologies.

Polyethylene glycol (PEG) deposited by means of PLD [6] was damaged even in case of finer deposition by IR laser, while PEG deposition by MAPLE from water matrix at 355 nm and 1064 nm was already successful [3, 8, 45]. PEG was deposited with no harm as a composite with poly(lactide-co-glycolide), which is suitable for controlled drug delivery from chloroform and dimethylsulfoxide (DMSO) at 266 nm [46-48], as well as a composite with metronidazole deposited at 248 nm from dichloromethane [49].

Similarly, metalocomplexes like acetyl acetonates used for sensors, as for example indium(III)acetylacetonate (Figure 12c), tin(IV)bis(acetylacetonate) dichloride and copper(II) para-tetratolylmethylporphyrine [50, 51] or matters for optical non-linear applications such as maleic anhydride–vinyl benzyl chloride, maleic anhydride–methyl methacrylate [52], were deposited successfully or just with minor modifications not affecting the function of the material. The MAPLE deposition proved itself to be effective at deposition of a wide range of biocompatible matters, such as biopolymers polycaprolactone [53], poly(D , L -lactide) [54, 55], poly(1,3-bis-(p-carboxyphenoxy propane)-co-sebacic anhydride [56], polysacharides as triacetate-pullulan [57], cinnamate-pullulan and tosylate-pullulan [58], proteins such as lysozyme [39, 59] and fibronectin [60].

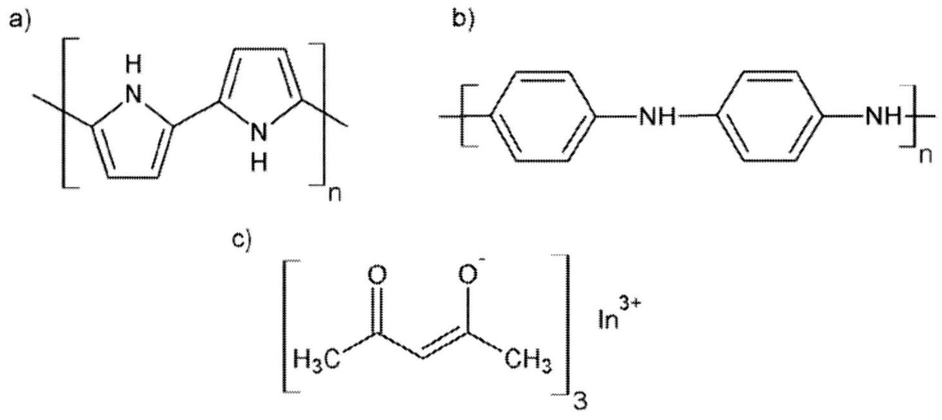

Figure 12. Examples of molecules deposited by MAPLE technology: (a) polypyrrole, (b) polyaniline, (c) indium(III)acetylacetonate.

An important group of matters that can be deposited using the MAPLE technology with no significant damage or loss of function are organic conductive polymers. To date, namely polyaniline (Figure 12b) [61], polypyrrole (Figure 12a) [42,62], poly(3-hexyl-thiophene) [63], poly [3-(4-octyloxyphenyl) thiophene] [64] and poly(3,4-ethylenedioxy-thiophene): poly(styrenesulfonate) [41] were successfully deposited.

A problem of the conductive polymers is generally their poor solubility in solvents, which is highly limiting for many technologies of deposition. The MAPLE technology also requires certain level of solubility of material in matrix; nevertheless, as mentioned above, it is possible to deposit from colloids or dispersions. Polyaniline was deposited from xylene and toluene matrix using the Nd:YAG laser of wavelength 266 and 355 nm with laser fluence up to 0,5 J per sq.cm.

Similarly successful deposition of polypyrrole was carried out from DMSO matrix with the KrF excimer laser on wavelength 192 and 248 nm, the ablation threshold for this polymer was determined to 0,3 J per sq.cm [42], poly(3-hexyl-thiophene) from orthoxylene again with the Nd:YAG laser at 266 nm. Slightly modified process was chosen by Johnson et al. [41], who used free-electron laser (FEL) tunable from 2 to 10 μm for deposition of composite poly(3,4-ethylenedioxy-thiophene):poly(styrenesulfonate). The laser was tuned to the absorption band of the solvent.

Such technology is using action of IR lasers onto vibration modes within the matrix molecule, which is able similarly to IR lasers in the RIR-PLD technology. Author is calling this technology a resonant infrared laser vapor deposition (RIR-LVD), within the terms used in this chapter we can however call it a RIR-MAPLE.

It is not always that the MAPLE technology is more efficient than classical PLD with suitably chosen parameters. Jelínek et al. [65] states, that bovine serum albumin (BSA) degrades at deposition with excimer KrF (248 nm) laser from DMSO matrix, even at relatively low fluences (0,3 J per sq.cm).

At the same time, DMSO can be considered to be a very good matrix at this wavelength [42]. On the other hand Sagawa et al. [27] has deposited BSA using the PLD technology with the IR Nd:YAG (1064 nm) laser at energies up to 5 J per sq.cm. The resulting film showed just slight signs of damage to the original material.

Direct Write Technologies – LIFT and MAPLE-DW

The LIFT or MAPLE-DW technologies are serial laser deposition technologies, whose interaction with the material is based on the same physical principle as the PLD or MAPLE technologies respectively. The main utilization of these technologies consists in the possibility to create various motives with high resolution (units of μm) on a surface, with no need for complicated lithographical procedures or mask usage. Therefore, they are sometime called “micropattering” or “laser writing” [66] and in many cases they may serve as an alternative to classical photolithography or the Ink-jet technology.

Arrangement of the LIFT deposition process is illustrated at Figure 13a. The basis consists of previously applied thin film (max. in hundreds of nm) of material on a material optically permeable for the wavelength of the used laser – together they form so-called ribbon.

Preparation of such thin film of the source material may be carried out even by other than laser deposition technique (typically, methods like spin-coating [67], dip-coating, plasma polymeration are used). Substrate, onto which the film is being deposited, is placed parallel to this film. In comparison with the parallel deposition methods, the distance between the ribbon and the substrate is shorter by several place-values; at PLD and MAPLE it reaches typically up to several tens of cm, at serial deposition methods this distance is counted in tens of μm. Such shortening leads to minimization of scattering of the cone of deposited particles of material and the resulting motive is finer. Further, the advantage of such arrangement is the possibility to work under normal atmospheric pressure with no need for complicated vacuum apparatuses.

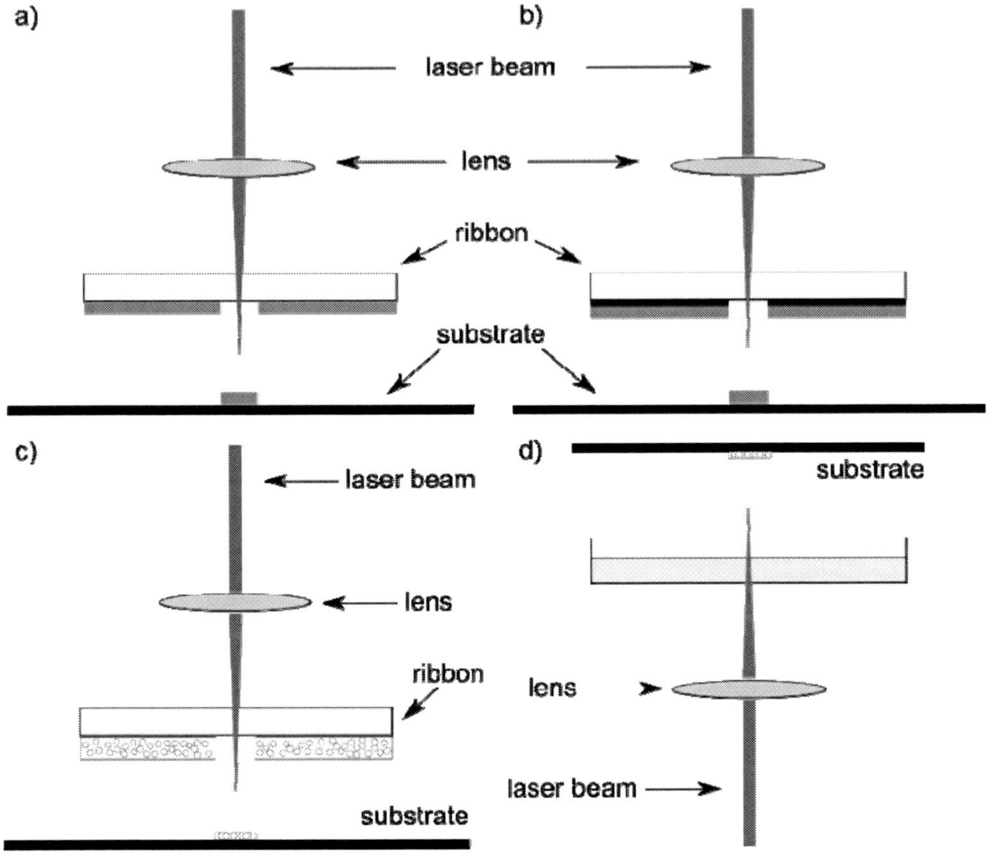

Figure 13. Arrangement of the deposition process of the direct write technologies.

The deposition apparatus includes usually also a PC-controlled xyz step motor, which enables to more the laser or the substrate in relation to the apparatus design and to draw a more complicated pattern.

The beam of a pulse or continuous laser is being focused with a lens. The focused beam is passing through an optically permeable pad (such as silicium) and it is absorbed on its boundary with a thin film of a source material. Material ablation would then cause formation of plasmatic cloud, which is accompanied by local pressure growth between the optically permeable pad and the thin film of source material. Consequently, part of the film gets torn off the ribbon and transferred onto the opposite substrate, where it adheres and forms a printed pixel.

Ideally, area of such pixel should correspond with the area of laser beam. A detailed view of the material transfer by means of direct write technologies is shown at Figure 14.

The ability of the LIFT technology to transfer brittle organic material non-destructively is directed by the same rules as for the PLD technology (see chapter Pulse Laser Deposition). Certain improvement in the ability of the LIFT technology in non-destructive transferring of the material was reached using so called Dynamic Release Layer method (DRL) [66], see

Figure 13b. Dynamic Release Layer forms an auxiliary intermediate layer between the optically permeable ribbon pad and the thin film of source material. It generally consists of a material absorbing at the wavelength of used laser that ables itself into a plasmatic cloud and such a way it improves the ability of the laser to print deposited material with no damage by laser radiation [66]. A significant disadvantage of DRL is that impurities may be introduced into the prepared films. Blister-based LIFT (BB-LIFT) may present a certain solution; this technology is using the intermediate film of sufficient thickness (up to units of μm), which prevents to full metal ablation. Deposited material is released only by local deformation of this metal film with no admixtures. For example Kononenko et al. [68] has transferred mesotetraphenylporphyrine non-destructively, using 50 to 1500 nm layer of titanium. A MAPLE-DW technology (see Figure 13c) is a modification of the LIFT technology, utilizing the ablation mechanism of the MAPLE technology. This deposition technology enables non-destructive transferring namely of macromolecules and polymer materials. Deposited material is frozen within a solution of matrix absorbing on the wavelength of used laser. Matrix is protecting the material from harmful laser radiation then it evaporates during deposition. Through collective collisions, the matrix is passing part of its energy to the deposited material that is consequently transferred onto substrate [69]. Another modification of direct write technology is a group of so-called film-free laser deposition, which enables printing of material from liquid samples. The material is transferred by a laser from thin surface film of liquid onto substrate. One of possible arrangement is shown at Figure 13d, another can be found for example here [70] or here [71]. The advantage of this technology is that there is no need to freeze the sample; the material is being applied in form of droplets similarly to the Ink-Jet technology.

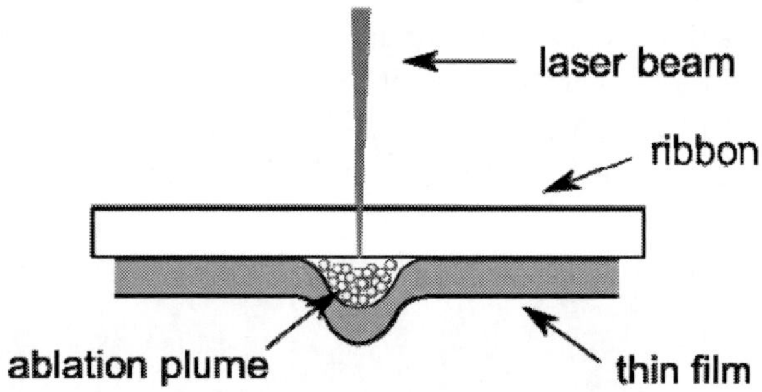

Figure 14. Material ablation during deposition by LIFT technology.

Deposition of Organic Materials Using the Direct Write Technology

Both above-mentioned technologies of direct writing have a wide range of applications, for example as technology for preparation of chemical sensors or sensitive transistor gates. Tsouti et al. [68] has successfully prepared active films of chemical sensors based on polyvinilpyrolidone, polyhydroxyethylmethacrylate, polyvinyl chloride, polymethymethacrylate, polyacrylic acid, poly(4-vinylpyridine) and polyvinyl alcohol using the LIFT technology. A 40 nm Cr layer was used as a DRL.

Cannata et al. [72] has prepared a pixel of size 40 x 40 sq.mm of polyepichlorohydrin (PECH), polyethyleneimine (PEI), and polyisobutylene (PIB) serving as a sensitive film of sensors, on the basis of Film Bulk Acoustic Resonators (FBARs). The experiment included also depositions of PECH, PEI and PIB using the MAPLE technology onto DRL consisting of dimethyl methyl phosphonate (DMMP).

Transistor gates on the basis of bis(2-phenylethynyl) end-substituted terthiophene were prepared by Rapp et al. [73]. These technologies are suitable for manipulation and production of sensor fields based on DNA [74, 75].

SUMMARY

Laser deposition technologies are interesting methods for preparation of thin defined films or fine patterns based on organic matters. This text dealt mainly with four main ones - Pulsed Laser Deposition, Matrix Assisted Pulsed Laser Evaporation, Laser Induced Forward Transfer, Matrix Assisted Pulsed Laser Evaporation – Direct Write. Each technology was described from the point of view of its principle and its ability to non-destructively transfer macromolecules, polymers and biologically functional matters. We have discussed here influences of various parameters onto deposited material and some fundamental results of depositing organic matters that have been reached within the research of these technologies.

Utilization of laser deposition technologies for deposition of organic matters is still perceived rather as an alternative to currently used simple methods, like for example the spin-coating or dip-coating. The main reasons are namely high expenses for purchasing the apparatus and demands for operator's experience. The main goal of this work, however, was to demonstrate that laser technologies might far exceed the currently used technologies by their precision and by their ability to transfer a wide range of materials usable in many branches of human activities. Defined films and structures do indeed offer even more comfortable conditions for further development.

We, the authors, hope that this text will be a source of inspiration for prospective persons interested in material engineering and deposition technologies and a tool for orientation in quite extensive sphere of laser depositions. For more detailed information on specific possibilities of described technologies please refer the enclosed list of institutes, which are engaged in them.

Table 2. Laser Workplace

State	Laser workplace
AUS	The Australian National University
	Laser Physics Centre, Research School of Physical Sciences and Engineering
AUT	Joanneum Research
	Laser Centre Leoben
AUT	Johannes-Kepler-Universität Linz
	Angewandte Physik
BGD	Bangladesh University of Engineering and Technology
	Department of Electrical and Electronic Engineering
CZE	Institute of Chemical Technology, Prague
	Department of Physics and Measurements
CZE	Academy of Science of the Czech Republic
	Institute of Physics

State	Laser workplace
DEU	University of Göttingen
	Institut für Materialphysik
DNK	Technical University of Denmark
	Risø National Laboratory
ESP	Institut de Ciència de Materials de Barcelona
FRA	Université Paris
	Groupe de Physique des Solides
GBR	University of Hull
	Department of Physics
GBR	University of Wales Swansea
	Department of Materials Engineering
GRC	Foundation for Research and Technology - Hellas
	Institute of Electronic Structure and Laser
GRC	National Center for Scientific Research Demokritos
	Institute of Microelectronics
GRC	National Technical University of Athens
	Physics Department
HKG	Hong Kong University of Science and Technology
	Department of Electrical Engineering
HKG	The Hong Kong Polytechnic University
HUN	University of Szeged
	Department of Optics and Quantum Electronics
CHN	Fudan University, Shanghai
	Laser Chemistry Institute
CHN	Nanjing University
	National Laboratory of Solid State Microstructures
IND	Indian Institute of Science
	Department of Materials Engineering
IND	Indian Institute of Technology Madras
	Department of Chemistry
IND	University of Pune
	Centre for Advanced Studies in Materials Science and Solid State Physics
ISR	Israel Institute of Metals
ITA	Consiglio nazionale delle ricerche
	Institute of inorganic methodologies and plasmas
ITA	Institute for Microelectronics and Microsystems
	IMM-CNR
ITA	Universita degli Studi di Napoli Federico II
	CNR-INFM Coherentia
ITA	Univerity of Salento
	Physics Department-L3 group
JPN	Keio University
	Faculty of Science and Technology
JPN	Kyoto Institute of Technology
	Department of Polymer Science and Engineering
KOR	Pusan National University
	Research Center for Dielectric and Advanced Matter Physics
KOR	Soonchunhyang University
	Department of Physics
POL	University of Rzeszow
	Institute of Physics
ROM	National Institute for Laser, Plasma and Radiation Physics
ROM	National Institute of Materials Physics
UKR	Institute of Semiconductor Physics
USA	Air Force Research Laboratory

Table 2. (Continued)

State	Laser workplace
USA	Duke University
	Department of Electrical and Computer Engineering
USA	Iowa State University
	Department of Mechanical Engineering
USA	Missouri State University
	Department of Physics, Astronomy, and Materials Science
USA	NASA Langley Research Center
USA	Naval Research Laboratory
USA	North Dakota State University
	Center for Nanoscale Science and Engineering
USA	Rensselaer Polytechnic Institute
USA	Seton Hall University
State	Laser workplace
USA	Thayer School of Engineering, Dartmouth College
USA	University of Missouri
	Department of Physics and Astronomy
USA	University of Virgina
	Department of Materials Science and Engineering
USA	Vanderbilt University
	W. M. Keck Foundation Free Electron Laser Laboratory
USA	Washington State University
	Department of Mechanical and Materials Engineering

ACKNOWLEDGMENTS

This work has been supported by the Grant Agency of Czech Republic projects No. P108/12/P802 and P108/11/1298 and from specific university research MSMT No. 21/2012.

REFERENCES

[1] D. Bauerle, *Appl. Surf. Sci.* 186 (1-4), 1-6 (2002).

[2] D. B. Chrisey, A. Pique, R. A. McGill, J. S. Horwitz, B. R. Ringeisen, D. M. Bubb and P. K. Wu, *Chem. Rev.* 103 (2), 553-576 (2003).

[3] A. Purice, J. Schou, P. Kingshottc, N. Pryds and M. Dinescu, *Appl. Surf. Sci.* 253 (19), 7952-7956 (2007).

[4] G. B. Blanchet and S. I. Shah, in *Appl. Phys. Lett.* (1993), Vol. 62, pp. 1026-1028.

[5] A. Pique, *Appl. Phys. a-Mater* 105 (3), 517-528 (2011).

[6] D. M. Bubb, A. O. Sezer, D. Harris, F. Rezae and S. P. Kelty, *Appl. Surf. Sci.* 253 (5), 2386-2392 (2006).

[7] G. Kecskemeti, T. Smausz, N. Kresz, Z. Toth, B. Hopp, D. Chrisey and O. Berkesi, *Appl. Surf. Sci.* 253 (3), 1185-1189 (2006).

[8] V. Califano, F. Bloisi, L. R. M. Vicari, P. Colombi, E. Bontempi and L. E. Depero, *Appl. Surf. Sci.* 254 (22), 7143-7148 (2008).

[9] A. Pique, R. C. Y. Auyeung, J. L. Stepnowski, D. W. Weir, C. B. Arnold, R. A. McGill and D. B. Chrisey, *Surf. Coat Tech.* 163, 293-299 (2003).

[10] J. M. Lackner, *Surf. Coat Tech.* 200 (5-6), 1439-1444 (2005).

[11] K. Grytsenko, O. Lytvyn, L. F. Ivanov, P. N. Grakovich, M. Sonntag and S. Schrader, *Appl. Surf. Sci.* 253 (19), 8028-8031 (2007).

[12] B. Losekrug, A. Meschede and H. U. Krebs, *Appl. Surf. Sci.* 254 (4), 1312-1315 (2007).

[13] B. Hopp, T. Smausz, G. Kecskemeti, A. Klini and Z. Bor, *Appl. Surf. Sci.* 253 (19), 7806-7809 (2007).

[14] G. Wisz, M. Kuzma, I. Virt, P. Sagan and I. Rudyj, *Appl. Surf. Sci.* 257 (12), 5319-5323 (2011).

[15] S. T. Li, E. Arenholz, J. Heitz and D. Bauerle, *Appl. Surf. Sci. 125* (1), 17-22 (1998).

[16] M. G. Norton, W. B. Jiang, J. T. Dickinson and K. W. Hipps, *Appl. Surf. Sci.* 96-8, 617-620 (1996).

[17] J. Blazevska-Gilev, Z. Bastl, J. Subrt, P. Stopka and J. Pola, *Polym. Degrad Stabil.* 94 (2), 196-200 (2009).

[18] W. A. Daoud, J. H. Xin, Y. H. Zhang and C. L. Mak, *Thin Solid Films* 515 (2), 835-837 (2006).

[19] G. B. Blanchet, C. R. Fincher, C. L. Jackson, S. I. Shah and K. H. Gardner, *Science* 262 (5134), 719-721 (1993).

[20] G. B. Blanchet, *Appl. Phys. Lett.* 62 (5), 479-481 (1993).

[21] Q. Luo, X. Y. Chen, Z. G. Liu, Z. M. Sun and N. B. Ming, *Appl. Surf. Sci.* 108 (1), 89-93 (1997).

[22] X. J. Yang, Y. X. Tang, M. Yu and Q. Z. Qin, *Thin Solid Films* 358 (1-2), 187-190 (2000).

[23] C. Hong, H. B. Chae, K. H. Lee, S. K. Ahn, C. K. Kim, T. W. Kim, N. I. Cho and S. O. Kim, *Thin Solid Films* 409 (1), 37-42 (2002).

[24] L. D. Wang and H. S. Kwok, Thin Solid Films 363 (1-2), 58-60 (2000).

[25] S. R. Farrar, A. E. A. Contoret, M. O'Neill, J. E. Nicholls, A. J. Eastwood, G. J. Richards, P. Vlachos and S. M. Kelly, *Appl. Surf. Sci.* 186 (1-4), 435-440 (2002).

[26] N. Cicco, T. Lopizzo, V. Marotta, A. Morone, M. Verrastro and V. Viggiano, *Appl. Surf. Sci.* 255 (10), 5458-5460 (2009).

[27] J. Sagawa, S. Nagare and M. Senna, *Appl. Surf. Sci.* 244 (1-4), 611-614 (2005).

[28] A. Palla-Papavlu, V. Dinca, V. Ion, A. Moldovan, B. Mitu, C. Luculescu and M. Dinescu, *Appl. Surf. Sci.* 257 (12), 5303-5307 (2011).

[29] Y. Tsuboi, N. Kimoto, M. Kabeshita and A. Itaya, *J. Photoch. Photobio A* 145 (3), 209-214 (2001).

[30] A. Das, S. Bera, M. Joseph, N. Sivakumar and A. Patnaik, *Appl. Surf. Sci.* 135 (1-4), 37-45 (1998).

[31] S. Kale, S. B. Ogale, J. P. Jog and V. M. Nadkarni, *Mater Lett.* 15 (4), 260-263 (1992).

[32] Y. T. Shin, S. W. Shin, J. Shin, K. Lee and M. Cha, *Thin Solid Films* 360 (1-2), 13-16 (2000).

[33] A. J. Salih, J. M. Marshall and J. M. Maud, *J. Non-Cryst Solids* 227, 1240-1244 (1998).

[34] R. Frycek, M. Vladimir, V. Martin, V. Filip, M. Jelinek and J. Nahlik, *Sensor Actuat B-Chem* 98 (2-3), 233-238 (2004).

[35] R. Cristescu, G. Socol, I. N. Mihailescu, M. Popescu, F. Sava, E. Ion, C. O. Morosanu and I. Stamatin, *Appl. Surf. Sci.* 208, 645-650 (2003).

[36] A. Pique, R. A. McGill, D. B. Chrisey, D. Leonhardt, T. E. Mslna, B. J. Spargo, J. H. Callahan, R. W. Vachet, R. Chung and M. A. Bucaro, *Thin Solid Films* 355, 536-541 (1999).
[37] A. P. Caricato, A. Luches, G. Leggieri, M. Martino and R. Rella, *Vacuum* 86 (6), 661-666 (2012).
[38] T. Tunno, A. P. Caricato, M. E. Caruso, A. Luches, M. Martino, F. Romano, D. Valerini and M. Anni, *Appl. Surf. Sci.* 253 (15), 6461-6464 (2007).
[39] A. Purice, J. Schou, P. Kingshott, N. Pryds and M. Dinescu, *Appl. Surf. Sci.* 253 (15), 6451-6455 (2007).
[40] C. N. Hunter, M. H. Check, J. E. Bultman and A. A. Voevodin, *Surf. Coat Tech.* 203 (3-4), 300-306 (2008).
[41] S. L. Johnson, H. K. Park and R. F. Haglund, *Appl. Surf. Sci.* 253 (15), 6430-6434 (2007).
[42] D. Kopecky, M. Vrnata, F. Vyslouzil, V. Myslik, P. Fitl, O. Ekrt, P. Matejka, M. Jelinek and T. Kocourek, *Thin Solid Films* 517 (6), 2083-2087 (2009).
[43] K. Rodrigo, P. Czuba, B. Toftmann, J. Schou and R. Pedrys, *Appl. Surf. Sci.* 252 (13), 4824-4828 (2006).
[44] A. T. Sellinger, A. H. Martin and J. M. Fitz-Gerald, *Thin Solid Films* 516 (18), 6033-6040 (2008).
[45] B. Toftmann, K. Rodrigo, J. Schou and R. Pedrys, *Appl. Surf. Sci.* 247 (1-4), 211-216 (2005).
[46] I. A. Paun, V. Ion, A. Moldovan and M. Dinescu, *Appl. Phys. Lett* 96 (24) (2010).
[47] I. A. Paun, V. Ion, A. Moldovan and M. Dinescu, *Appl. Surf. Sci.* 257 (12), 5259-5264 (2011).
[48] I. A. Paun, V. Ion, A. Moldovan and M. Dinescu, *Appl. Phys. a-Mater* 106 (1), 197-205 (2012).
[49] R. Cristescu, C. Popescu, A. Popescu, S. Grigorescu, I. N. Mihailescu, D. Mihaiescu, S. D. Gittard, R. J. Narayan, T. Buruiana, I. Stamatin and D. B. Chrisey, *Appl. Surf. Sci.* 255 (24), 9873-9876 (2009).
[50] R. Frycek, M. Jelinek, T. Kocourek, P. Fitl, M. Vrnata, V. Myslik and M. Vrbova, *Thin Solid Films* 495 (1-2), 308-311 (2006).
[51] R. Frycek, F. Vyslouzil, V. Myslik, M. Vrnata, D. Kopecky, O. Ekrt, P. Fitl, M. Jelinek, T. Kocourek and R. Sipula, *Sensor Actuat B-Chem* 125 (1), 189-194 (2007).
[52] G. Socol, I. N. Mihailescu, A. M. Albu, S. Antohe, F. Stanculescu, A. Stanculescu, L. Mihut, N. Preda, M. Socol and O. Rasoga, *Appl. Surf. Sci.* 255 (10), 5611-5614 (2009).
[53] R. Cristescu, A. Doraiswamy, G. Socol, S. Grigorescu, E. Axente, D. Mihaiescu, A. Moldovan, R. J. Narayan, I. Stamatin, I. N. Mihailescu, B. J. Chisholm and D. B. Chrisey, *Appl. Surf. Sci.* 253 (15), 6476-6479 (2007).
[54] R. Cristescu, A. Doraiswamy, T. Patz, G. Socol, S. Grigorescu, E. Axente, F. Sima, R. J. Narayan, D. Mihaiescu, A. Moldovan, I. Stamatin, I. N. Mihailescu, B. Chisholm and D. B. Chrisey, *Appl. Surf. Sci.* 253 (19), 7702-7706 (2007).
[55] T. M. Patz, A. Doraiswamy, R. J. Narayan, N. Menegazzo, C. Kranz, B. Mizaikoff, Y. Zhong, R. Bellamkonda, J. D. Bumgardner, S. H. Elder, X. F. Walboomers, R. Modi and D. B. Chrisey, *Mat. Sci. Eng. C-Bio S* 27 (3), 514-522 (2007).

[56] R. Cristescu, C. Popescu, G. Socol, A. Visan, I. N. Mihailescu, S. D. Gittard, P. R. Miller, T. N. Martin, R. J. Narayan, A. Andronie, I. Stamatin and D. B. Chrisey, *Appl. Surf. Sci.* 257 (12), 5287-5292 (2011).

[57] R. Cristescu, G. Dorcioman, C. Ristoscu, E. Axente, S. Grigorescu, A. Moldovan, I. N. Mihailescu, T. Kocourek, M. Jelinek, M. Albulescu, T. Buruiana, D. Mihaiescu, I. Stamatin and D. B. Chrisey, *Appl. Surf. Sci.* 252 (13), 4647-4651 (2006).

[58] M. Jelinek, R. Cristescu, E. Axente, T. Kocourek, J. Dybal, J. Remsa, J. Plestil, D. Mihaiescu, M. Albulescu, T. Buruiana, I. Stamatin, I. N. Mihailescu and D. B. Chrisey, *Appl. Surf. Sci.* 253 (19), 7755-7760 (2007).

[59] A. Purice, J. Schou, N. Pryds, A. Filipescu and A. Dinescu, *Appl. Surf. Sci.* 254 (4), 1244-1248 (2007).

[60] F. Sima, P. Davidson, E. Pauthe, L. E. Sima, O. Gallet, I. N. Mihailescu and K. Anselme, *Acta Biomater* 7 (10), 3780-3788 (2011).

[61] C. Constantinescu, N. Scarisoreanu, A. Moldovan, M. Dinescu and C. Vasiliu, *Appl. Surf. Sci.* 253 (19), 7711-7714 (2007).

[62] A. Pique, P. Wu, B. R. Ringeisen, D. M. Bubb, J. S. Melinger, R. A. McGill and D. B. Chrisey, *Appl. Surf. Sci.* 186 (1-4), 408-415 (2002).

[63] A. Gutierrez-Llorente, G. Horowitz, R. Perez-Casero, J. Perriere, J. L. Fave, A. Yassar and C. Sant, *Org. Electron* 5 (1-3), 29-34 (2004).

[64] F. Bloisi, A. Cassinese, R. Papa, L. Vicari and V. Califano, *Thin Solid Films* 516 (7), 1594-1598 (2008).

[65] M. Jelinek, J. Remsa, E. Brynda, A. Houska and T. Kocourek, *Appl. Surf. Sci.* 254 (4), 1240-1243 (2007).

[66] I. Zergioti, M. Makrygianni, P. Dimitrakis, P. Normand and S. Chatzandroulis, *Appl. Surf. Sci.* 257 (12), 5148-5151 (2011).

[67] V. Dinca, A. Palla-Papavlu, I. Paraico, T. Lippert, A. Wokaun and M. Dinescu, *Appl. Surf. Sci.* 257 (12), 5250-5254 (2011).

[68] T. V. Kononenko, I. A. Nagovitsyn, G. K. Chudinova and I. N. Mihailescu, *Appl. Surf. Sci.* 256 (9), 2803-2808 (2010).

[69] D. B. Chrisey, A. Pique, J. Fitz-Gerald, R. C. Y. Auyeung, R. A. McGill, H. D. Wu and M. Duignan, *Appl. Surf. Sci.* 154, 593-600 (2000).

[70] A. Patrascioiu, M. Duocastella, J. M. Fernandez-Pradas, J. L. Morenza and P. Serra, *Appl. Surf. Sci.* 257 (12), 5190-5194 (2011).

[71] A. Patrascioiu, J.M. Fernández-Pradas, J.L. Morenza, P. Serra, *Appl. Surf. Sci.* (2011), in press.

[72] D. Cannata, M. Benetti, F. Pietrantonio, E. Verona, A. Palla-Papavlu, V. Dinca, M. Dinescu, T. Lippert, *Sens Actuators B: Chem*(2012), in press.

[73] L. Rapp, F. Serein-Spirau, J.-P.Lere-Porte, A. P. Alloncle, P.Delaporte,F.Fages, C. Videlot-Ackermann, *Org Electron* (2012), in press.

[74] M. Colina, P. Serra, J. M. Fernandez-Pradas, L. Sevilla and J. L. Morenza, *Biosens Bioelectron* 20 (8), 1638-1642 (2005).

[75] G. Tsekenis, M. Chatzipetrou, J. Tanner, S. Chatzandroulis, D. Thanos, D. Tsoukalas, I. Zergioti, *SensActuators B: Chem*(2012), in press.

INDEX

A

B

C

D

E

F

G

H

I

K

L

M

N

O

P

Q

R

U

V

W

Y